ADVANCES IN
ANTIVIRAL DRUG DESIGN

Volume 3 • 1999

ADVANCES IN ANTIVIRAL DRUG DESIGN

Editor: E. DE CLERCQ
Rega Institute for Medical Research
Katholieke Universiteit Leuven
Leuven, Belgium

VOLUME 3 • 1999

JAI PRESS INC.
Stamford, Connecticut

CONTENTS

LIST OF CONTRIBUTORS

Murty N. Arimilli   Gilead Sciences, Inc.
Foster City, California

Masanori Baba   Center for Chronic Viral Diseases
Kagoshima University
Kagoshima, Japan

Norbert Bischofberger   Gilead Sciences, Inc.
Foster City, California

Gary J. Bridger   AnorMED Inc.
Langley, British Columbia, Canada

Kenneth C. Cundy   Gilead Sciences, Inc.
Foster City, California

Jeff C. Dyason   Department of Medicinal Chemistry
Monash University
Parkville, Victoria, Australia

E. De Clercq   Rega Institute for Medical Research
Katholieke Universiteit Leuven
Leuven, Belgium

Joseph P. Dougherty   Gilead Sciences, Inc.
Foster City, California

Jean-Christophe G. Graciet   Atlanta Veterans Affairs Medical
Decatur, Georgia, and
Department of Pediatrics
Emory University School of Medicine
Atlanta, Georgia

Kazuhiro Haraguchi   School of Pharmaceutical Sciences
Showa University
Tokyo, Japan

Hiroyuki Hayakawa   School of Pharmaceutical Sciences
                             Showa University
                             Tokyo, Japan

Tadashi Miyasaka   School of Pharmaceutical Sciences
                             Showa University
                             Tokyo, Japan

Raymond F. Schinazi   Atlanta Veterans Affairs Medical Center
                             Decatur, Georgia, and
                             Department of Pediatrics
                             Emory University School of Medicine
                             Atlanta, Georgia

Renato T. Skerlj   AnorMED Inc.
                             Langley, British Columbia, Canada

David K. Stammers   Laboratory of Molecular Biophysics
                             Oxford University
                             Oxford, England

David I. Stuart   Laboratory of Molecular Biophysics
                             Oxford University
                             Oxford, England

Hiromichi Tanaka   School of Pharmaceutical Sciences
                             Showa University
                             Tokyo, Japan

Mark von Itzstein   Department of Medicinal Chemistry
                             Monash University
                             Parkville, Victoria, Australia

Richard T. Walker   School of Chemistry
                             University of Birmingham
                             Birmingham, England

PREFACE

This represents the third volume of *Advances in Antiviral Drug Design*. It continues the tradition of the preceding volumes published in 1993 (Volume 1) and 1996 (Volume 2). These volumes dealt with (i) uncoating inhibitors for picornavirus infections, (ii) broad-spectrum antiviral nucleoside analogs (like ribavirin), (iii) acyclic nucleoside analogs (like acyclovir and ganciclovir), (iv) acyclic nucleoside phosphonates (like cidofovir and adefovir), (v) dideoxynucleoside analogs (such as zidovudine, didanosine, zalcitabine, and stavudine), (vi) antisense oligonucleotides, (vii) *S*-adenosylhomocysteine (AdoHcy) hydrolase inhibitors, (viii) carbocyclic nucleoside analogs, (ix) nucleotide prodrugs (bypassing the initial phosphorylation step), and (x) HIV protease inhibitors.

Meanwhile, the antiviral drug armamentarium has been growing at an unabated pace; it now includes more than 30 compounds, half of which are for the treatment of human immunodeficiency virus (HIV) infections. In recent years, we have witnessed the advent of seven NRTIs (nucleoside/nucleotide reverse transcriptase inhibitors: zidovudine, didanosine, zalcitabine, stavudine, lamivudine, abacavir, and adefovir dipivoxil), three NNRTIs (nonnucleoside reverse transcriptase inhibitors: nevirapine, de-

lavirdine, and efavirenz), and five HIV protease inhibitors (saquinavir, ritonavir, indinavir, nelfnavir and amprenavir). Others are forthcoming.

The present volume of *Advances in Antiviral Drug Design* is keeping up
with the recent progress made in the field of antiviral drug research and
highlights five specific directions that have opened new avenues for the
treatment of virus infections.

First, the use of lamivudine (3TC) for the treatment of HIV infections, and
its more recent introduction for the treatment of hepatitis B virus (HBV)
infections, has heralded the transition of D- to L-nucleosides in the antiviral
nucleoside drug design, and it is likely that the future will provide more
nucleosides of the L-configuration, such as (–)FFC (emtricitabine) and
L-FMAU, as will be described by J.-C.G. Graciet and R.F. Schinazi.

Second, the acyclic purine nucleoside phosphonates, i.e. PMEA (adefovir)
and PMPA (tenofovir), offer great potential as both anti-HIV and anti-HBV
agents, and both compounds have been the subject of advanced clinical trials
in their oral prodrug form (adefovir dipivoxil and tenofovir disoproxyl), as
mentioned by M.N. Arimilli, J.P. Dougherty, K.C. Cundy, and N. Bischofberger.

Third, with the advent of nevirapine, delavirdine, and efavirenz, the
NNRTIs have definitely come of age. Emivirine (MKC-442), a derivative of
the original HEPT analog that was described in 1989 has now proceeded
through pivotal clinical studies, and how this class of compounds evolved is
presented in the account of H. Tanaka and his colleagues.

Fourth, at the end of 1999, anticipating on the next winter influenza
offensive, we should have at end two compounds that specifically inhibit
influenza A and B virus infections: zanamivir (by the intranasal route) and
oseltamivir (by the oral route). Both compounds have proved effective in the
prophylaxis and treatment of influenza A and B virus infections and act
through the same mechanism; that is by blocking the viral neuraminidase (or
sialidase), a key enzyme that allows the virus to spread from one cell to
another (within the respiratory mucosal tract). The design of these sialidase
inhibitors will be presented by M. von Itzstein and J.C. Dyason.

Fifth, the discovery (in 1996) of the chemokine receptors CXCR4 and
CCR5 as essential coreceptors (in addition to the CD4 recjeptor) for HIV
entry into the cells, has boosted an enormous interest in potential antagonists
of these receptors. The bicyclams represent the first low-molecular-weight
compounds targeted at CXCR4, the coreceptor used by the more pathogenic,
T-lymphotropic, HIV strains, to enter the cells. They will be addressed by
G.J. Bridger and R.T. Skerlj.

The five topics covered in this third volume of *Advances in Antiviral Drug
Design* are in the front line of the present endeavors towards the chemother-

apy of virus infections. They pertain to the combat against three of the most important virus infections of current times: HIV, HBV, and influenza virus.

E. De Clercq
Editor

FROM D- TO L-NUCLEOSIDE ANALOGS AS ANTIVIRAL AGENTS

Jean-Christophe G. Graciet and
Raymond F. Schinazi

Advances in Antiviral Drug Design
Volume 3, pages 1–68.
Copyright © 1999 by JAI Press Inc.
All rights of reproduction in any form reserved.
ISBN: 0-7623-0201-1

I. INTRODUCTION

Since the discovery of β-D-5-iodo-2′-deoxyuridine as a potent anti-herpetic drug for the treatment of herpes keratitis in the 1960s by Prusoff,[1,2] a host of nucleosides and non-nucleosides antiviral agents have been synthesized and developed. However, it was not until the late 1980s that L-nucleosides were recognized as agents with great promise in the treatment of HIV and subsequently HBV infection.[3-7] The FDA approval of (−)-β-2′,3′-dideoxy-3′-thiacytidine for the treatment of HIV in 1996, especially when used in combined regimens, was an important milestone in nucleosides antiviral chemotherapy. Since then, intensive studies on unnatural L-nucleosides and enriched enantiomers of well-established D-compounds have been conducted, leading to a flurry of exciting new agents with activity against HIV, HBV, and herpesviruses, including EBV. Some of the less selective L-nucleosides are also being considered as anticancer agents.[8] More recently, we proposed the use of racemic compounds as antiviral agents, provided they have no antagonism and toxicity, since they would have different resistance profiles.[9] This review examines the various successful synthetic approaches to enantiomeric unnatural nucleosides. Whenever possible, the rationale for making these compounds in addition to their potential use as antiviral and anticancer agents, as well as plant chemotactic agents, such as L-adenosine[10] is included.

II. L-NUCLEOSIDES

A. L-Ribonucleosides

Holý et al.[11] reported the synthesis of a series of β-L-ribonucleosides such as L-cytidine (**1**), L-uridine (**2**), L-thymidine (**3**), L-guanosine (**4**),

L-adenosine (**5**), and L-6-azauridine (**6**) (Figure 1). Burnstock et al.[12] also obtained the 2-methylthio-β-L-adenosine (**7**) (Figure 1).

Starting from the benzyl-β-L-arabinopyranoside, Holý et al.[11] prepared the key intermediate **8** in five steps. The nucleosidic bond was formed using a reactive 1,2-epoxide **9** and various sodium salt of nucleic bases in DMF (Scheme 1). Deprotection with 80% aqueous acetic acid at reflux for the cytosine, thymine, uracil, guanine, and adenine nucleosides or hydrogenolysis in presence of palladium in AcOH/HCl for the N^3-benzyl-6-azauracil nucleoside, provided the corresponding β-L-ribonucleosides (**1–6**, Figure 1).

Using a methodology developed by Gough et al.,[13] Cusack et al.[14] synthesized the 2-chloro- and 2-azido-β-L-adenosine (**11** and **12**, respectively, Scheme 2). By fusion of the 2,6-dichloropurine to the β-L isomer of the 1-O-acetyl-2,3,5-tri-O-benzoyl-ribofuranose **10** at 155 °C, followed by treatment with methanolic ammonia, the 2-chloro-β-L-adenosine **11** was obtained. Further treatment with hydrazine and nitrous acid afforded the 2-azido-β-L-adenosine **12**, as described by Schaeffer et al.[15] (Scheme 2).

Asai et al.[16] have previously described the synthesis of β-L-purine-ribonucleosides by optical resolution of D,L-isomers using microorganisms. The synthesis of the D,L-mixture of 1-O-acetyl-2,3,5-tri-O-benzoyl-ribofuranose **10** was accomplished starting from the D,L-ribose. After replacement of the 1-O-acetyl group by chloride (**13**), the nucleoside derivative was obtained by condensation of the sugar-halide and the chloromercuric-6-benzamidopurine using the Davoll and Lowy method.[17] Removal of the protecting groups using saturated methanolic ammonia solution provided the D,L-adenosine mixture (**14**) (Scheme 3).

Incubation of β-D,L-adenosine **14** for 48 h in a cell suspension of a strain *Pseudomonas Ovalis* PS-264 caused total enzymatic decomposition of the β-D-adenosine and provided the pure β-L-adenosine **5** and after deamination, pure β-L-inosine **15**.[16]

Similarly, Asai et al.[16] used snake venom for the resolution of a mixture of D,L-adenosine monophosphate **19** (D,L-AMP) and D,L-inosine monophosphate **20** (D,L-IMP, Scheme 4). Starting from the D,L-adenosine **14** which was converted in D,L-inosine **16** using $NaNO_2$ in acetic acid, Asai et al first protected the 2,3-diol as an isopropylidene derivative using acetone under acidic conditions (**17** and **18**, respectively,

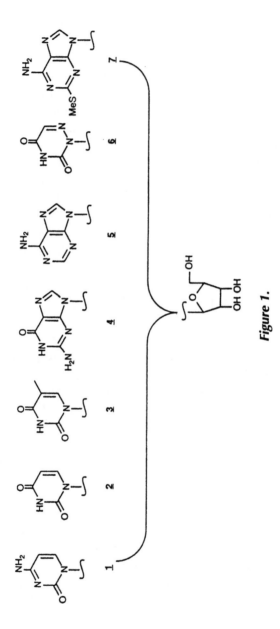

Figure 1.

4

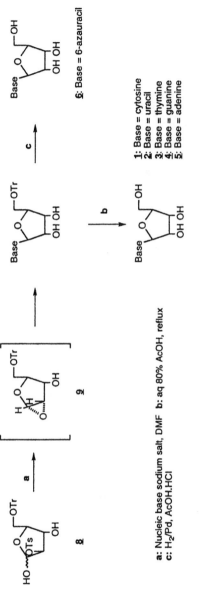

6: Base = 6-azauracil

1: Base = cytosine
2: Base = uracil
3: Base = thymine
4: Base = guanine
5: Base = adenine

a: Nucleic base sodium salt, DMF   b: aq 80% AcOH, reflux
c: H₂/Pd, AcOH.HCl

Scheme 1.

5

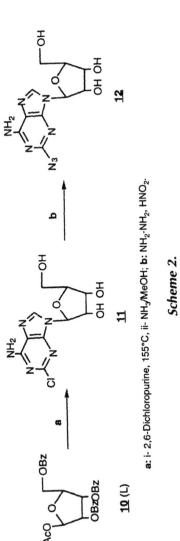

a: i- 2,6-Dichloropurine, 155°C, ii- NH₃/MeOH; **b:** NH₂-NH₂, HNO₂.

Scheme 2.

6

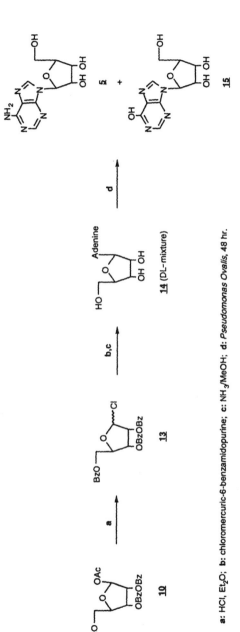

Scheme 3.

a: HCl, Et$_2$O; **b:** chloromercuric-6-benzamidopurine; **c:** NH$_3$/MeOH; **d:** *Pseudomonas Ovalis*, 48 hr.

7

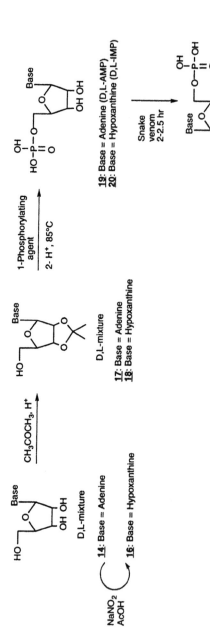

Scheme 4.

8

Scheme 4). Using phenylphosphorodichloridate for the D,L-adenosine and tetra-*p*-nitrophenyl pyrophosphate for D,L-inosine, followed in both cases by removal of the protecting group using acidic hydrolysis, the corresponding D,L-adenosine and inosine 5′-monophosphate were obtained (**19** and **20,** respectively, Scheme 4). Incubation of these mixtures with the snake venom for 2 h led to the selective 5′-dephosphorylation of the D-enantiomer of adenosine and inosine, leaving unreacted the L-adenosine monophosphate **21** (L-AMP) and L-inosine monophosphate **22** (L-IMP, Scheme 4).

Génu-Dellac et al.[18] synthesized the α-isomer of the L-ribo-furanosylthymidine (**26**, Scheme 5). Starting from the 3,5-di-*O*-benzoyl-α-L-arabinothymidine **23**,[19] Génu-Dellac et al. proceeded with the elimination of the 2′-hydroxyl group using phenylthionochloride in presence of DMAP followed by reaction with tributyltin hydride with AIBN. Deprotection of the 2′-deoxy compound using methanolic ammonia afforded the α-L-2′-deoxythymidine **24** (Scheme 5). Protection of the 5′-hydroxyl group using monomethoxytrityl chloride in pyridine followed by reaction of DAST in pyridine and then trifluoroactic acid provided the α-L-2′,3′-dideoxy-2′,3′-didehydrothymidine **25**. Treatment of **25** with OsO_4 yielded to the α-L-thymidine **27** and to the α-L-lyxofuranosylthymine **26** (Scheme 5).

B. Miscellaneous L-Nucleosides

Aside from L-ribonucleosides, there has been interest in other furanose rings as sugar moiety, such as arabino-, xylo-, and lyxofuranose. Holý[20] described the synthesis of pyrimidine β-L-arabinonucleosides such as L-arabinouridine (**28**) (Figure 2) using alkaline hydrolysis of an 2,2′-anhydro-L-uridine intermediate whose synthesis is described later (Scheme 9). Jansons et al.[21] reported the synthesis of a series of α-L-purine-arabinonucleosides (**29–40**) (Figure 2). Génu-Dellac et al.[19] also prepared a series of α-L-arabinonucleosides with the five naturally occurring nucleic acid bases (**29, 41–44,** Figure 2).

Starting from N^6-benzoyladenine, Jansons et al.[21] obtained the nucleoside derivatives by condensation on a 1,2,3,5-tetra-*O*-acetyl-L-arabinofuranose using tin(IV) chloride (Scheme 6). After deblocking

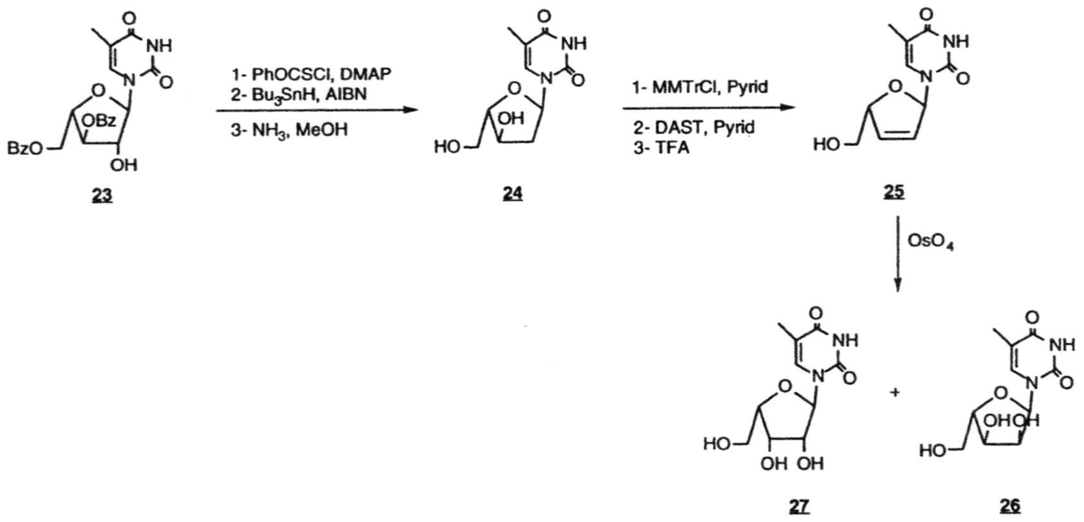

Scheme 5.

Reaction conditions shown in scheme:

23 → 24: 1- PhOCSCl, DMAP; 2- Bu₃SnH, AIBN; 3- NH₃, MeOH

24 → 25: 1- MMTrCl, Pyrid; 2- DAST, Pyrid; 3- TFA

25 → 26 + 27: OsO₄

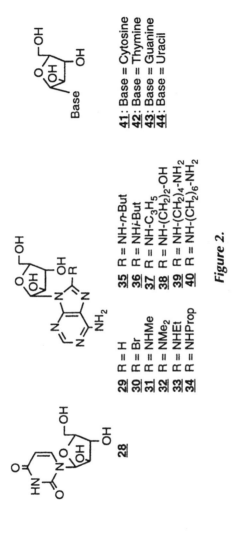

28

29 R = H
30 R = Br
31 R = NHMe
32 R = NMe₂
33 R = NHEt
34 R = NHProp

35 R = NH-*n*-But
36 R = NH*i*-But
37 R = NH-C₃H₅
38 R = NH-(CH₂)₂-OH
39 R = NH-(CH₂)₄-NH₂
40 R = NH-(CH₂)₆-NH₂

41: Base = Cytosine
42: Base = Thymine
43: Base = Guanine
44: Base = Uracil

Figure 2.

11

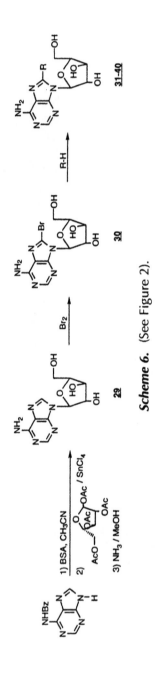

Scheme 6. (See Figure 2).

12

the amino group by methanolic ammonia solution (**29**), the introduction on C8 of the adenine base was achieved by bromination of this position (**30**), followed by displacement by various aminoalkyl groups (**31–40**, Scheme 6).

The synthesis of β-L-xylofuranosyl purines and pyrimidines nucleosides was reported by Gosselin et al.[22] Starting from the L-xylose, the 1,2- and 3,5-diacetonide derivative **45** was obtained, which was 3,5-deprotected and 3,5-dibenzoylated (**46**) using successively HCl, NaHCO₃, and benzoylchloride in pyridine/chloroform. Simultaneous deprotection and acetylation of the 1,2-acetonide **46** using acetic acid and acetic anhydride in sulfuric acid provided the 1,2-di-*O*-acetyl-3,5-*O*-dibenzoyl-L-xylofuranosyl derivative **47**. Using either tin chloride or TMSOTf, the condensation of the five naturally occurring nucleic bases was then achieved and a final deprotection provided the six desired nucleoside analogs (**48–53**, Scheme 7).

Proceeding with the multistep synthesis previously described by Holý[20], Lin et al.[23] obtained several derivatives of the β-L-arauridine **28** from L-arabinose **54** (Scheme 8).

Iodination of β-L-arauridine **28** was achieved with silver trifluoroacetate and iodine in dry dioxane, producing the corresponding 5-iodo-β-L-arauridine **57** in good yield. Reaction of **57** with methylacrylate, bis(acetato)bis(triphenylphosphine)palladium(II), and triethylamine-dioxane afforded the β-L-5-(*E*)-methoxycarbonylvinyl-arauridine **58**. Hydrolysis with sodium hydroxide, followed by acidification and reaction with *N*-bromosuccinimide in dry DMF in presence of KHCO₃, provided the β-L-5-(*E*)-bromovinyl-arauridine **59** (Scheme 8).

Holý et al.[11] also reported the synthesis of the β-L-lyxouridine **63** by conversion of the β-L-uridine **2** (Scheme 9). After *O*-mesylation of the 2′-, 3′- and 5′-hydroxy groups (**60**), basic hydrolysis using sodium hydroxide afforded the 2,2′-anhydro derivative **61** in which the 5′-*O*-mesyl group was then replaced by a benzoyl group in compound **62**. Final deprotection using NH₃/MeOH provided the β-L-lyxouridine **63** (Scheme 9).

Génu-Dellac et al.[19] prepared the α-L-lyxofuranosylthymine **26** as already described in Scheme 5.

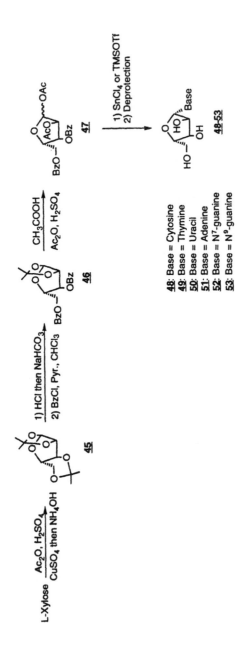

Scheme 7.

14

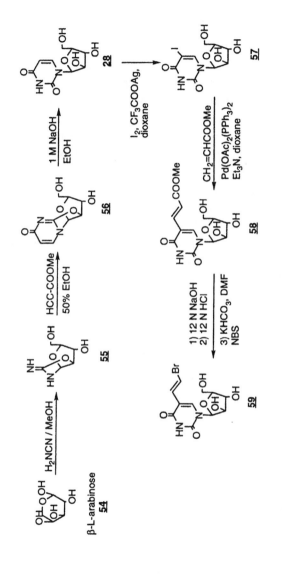

Scheme 8.

15

III. THE 2'- OR 3'-DEOXY-L-NUCLEOSIDES

A. 2'-Deoxy-L-nucleosides and Their Derivatives

Starting from the 2,2'-*O*-anhydro-L-uridine **56** (Scheme 8), Holý prepared a series of 2'-deoxy-β-L-pyrimidine nucleoside analogs such as 2'-deoxy-β-L-cytidine (**64**),[20] 2'-deoxy-β-L-thymidine (**65**),[20] 2'-deoxy-β-L-uridine (**66**),[20,24] 2'-deoxy-5-bromo-β-L-uridine (**67**),[20] and 2'-deoxy-5-hydroxymethyl-β-L-uridine (**68**)[20] (Scheme 10).

Reaction of **56** with benzoyl chloride in pyridine followed by treatment with HCl or LiI afforded compounds **68–70**. Treatment with tributyltin hydride led to the 3',5'-*O*-dibenzoyl-2'-deoxy-L-uridine **71**, which yielded the 2'-deoxy-L-cytidine **64** when it was treated with phosphorus pentasulfide and then methanolic ammonia. When compound **71** was directly treated with methanolic ammonia, 2'-deoxy-L-uridine **66** was obtained. Bromination of **66** afforded the 5-bromo-2'-deoxy-L-uridine **67**, while treatment of **66** with formaldehyde with NaOH led to 5-hydroxymethyl-2'-deoxy-L-uridine **68**. Eth-

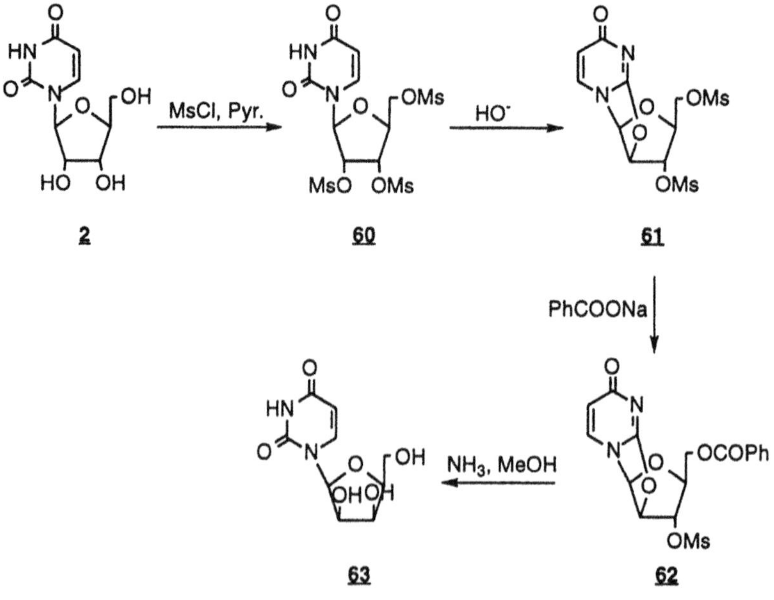

Scheme 9.

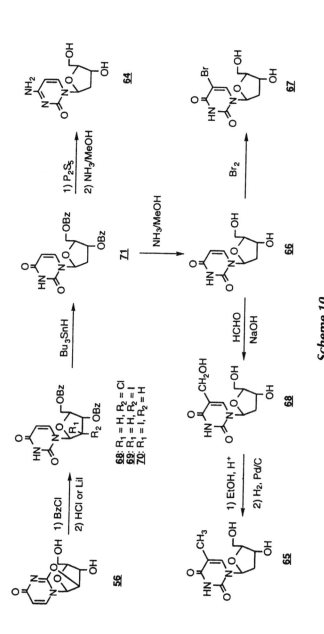

Scheme 10.

erification with ethanol in acidic conditions followed by hydrogenation on Pd/C gave 2′-deoxy-L-thymidine 65.

Robins et al.[25] reported the synthesis of α- and β-2′-deoxy-L-purine nucleoside analogs. Using a procedure described by Hoffer,[26] the free 2-deoxy-L-ribose, prepared from 3,5-di-*O*-acetyl-L-arabinol,[27] was converted to the 1-*O*-methyl-3,5-di-*O*-*p*-toluyl-2-deoxy-L-*erythro*-pentofuranose 72. Hydrolysis with diluted acid and further acetylation of 72 afforded the 1-*O*-acetyl-3,5-di-*O*-*p*-toluyl-2-deoxy-L-*erythro*-pentofuranose 73 (Scheme 11). Fusion[28,29] of 72 with 2,6-dichloropurine at 143 °C in dichloroacetic acid and treatment with methanolic ammonia led to 2-chloro-6-amino-9-(2-deoxy-α-L-*erythro*-pentofuranosyl)purine 74 and 2-chloro-6-amino-9-(2-deoxy-β-L-*erythro*-pentofuranosyl)purine 75. Further treatment with concentrated aqueous ammonia and hydrogenation on Pd/C afforded the 2′-deoxy-α-L-adenosine 76 and the 2′-deoxy-β-L-adenosine 77 (Scheme 11). Fusion of 73 with 2-fluoro-5-benzyloxypurine at 155 °C in dichloroacetic acid *in vacuo* and treatment with NH₃/MeOH at 85

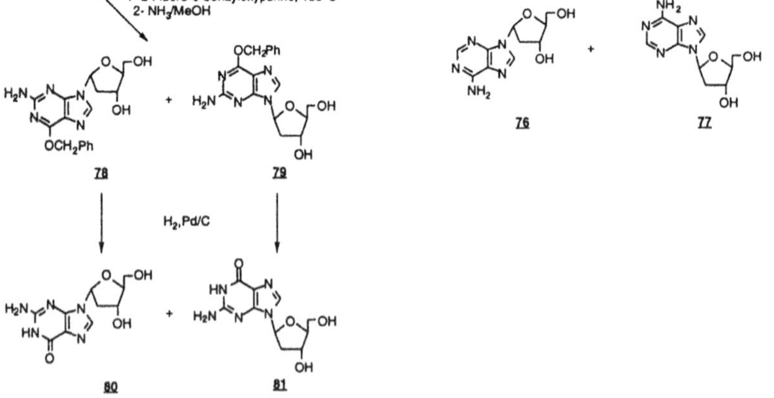

Scheme 11.

°C provided the 2′-deoxy-2-amino-6-benzyloxy-α-L-purine **78** and the 2′-deoxy-2-amino-6-benzyloxy-β-L-purine **79**. Final hydrogenation with Pd/C afforded the 2′-deoxy-α-L-guanosine **80** and the 2′-deoxy-β-L-guanosine **81** (Scheme 11).

More recently, Spadari et al.[30] reported the synthesis of the 2′-deoxy-β-L-nucleosides of the five naturally occurring nucleic acid bases (**64–66, 77** and **81**) using the procedures developed by Holý[20] and Robins et al.[25]

Several studies focused on the introduction of a substituent such as fluorine at this 2′-position of L-nucleoside analogs in order to increase the stability of the glycosidic bond.[31,32] Some of the 2′-fluoro analogs resulted in compounds with a broad antiviral activity against different herpesviruses.[33]

Xiang et al.[34] prepared some 2′-deoxy-β-L-ribofuranosyl-pyrimidine and -purine nucleosides in which the 2′-position is disubstituted by two fluorines. Starting from the L-gulonic-γ-lactone, they synthesized the 2,2-difluoro-intermediate **82** in four steps. After protection of hydroxyl groups using *tert*-butyldimethylsilyl chloride in DMF and DIBAL-H reduction, the key intermediate **83** was obtained. Further mesylation using mesyl chloride provided the intermediate **84** (Scheme 12). Reaction of **83** with silylated base in the presence of NaI in CH_3CN, followed by deprotection using sodium methoxide in methanol and then *n*-Bu_4NF in THF, provided the desired 2′-deoxy-2′,2″-difluoro-pyrimidine nucleosides as thymine analogs **85** and **86** and cytosine derivatives **87** and **88** (Scheme 12). Condensation of **82** with 6-chloropurine using Mitsunobu conditions and reaction with *n*-Bu_4NF in THF provided the 6-chloropurine derivatives **89** and **90**. Further reaction using methanolic ammonia yielded to the 2′-deoxy-2′,2″-difluoro-L-adenosine analogs **91** and **92** (Scheme 12).

Ma et al.[35] reported the synthesis of the 2′-fluoro-5-methyl-β-L-arabinofuranosyluracil (L-FMAU, **95**) and related thymine and cytosine derivatives (**96–106**). Starting from the 1,3,5-tri-*O*-benzoyl-2-fluoro-α-L-arabinofuranose **93**, as already described by Chu et al.,[36] the condensation step was achieved using first hydrogen bromide in acetic acid at room temperature leading to intermediate **94,** which was immediately coupled to the silylated nucleic bases in 1,2-dichloroethane at reflux. Following removal of benzoyl groups

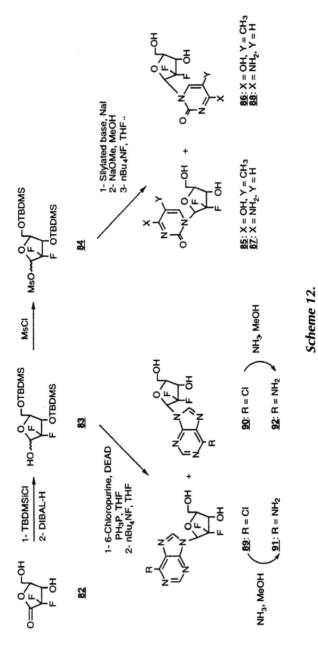

Scheme 12.

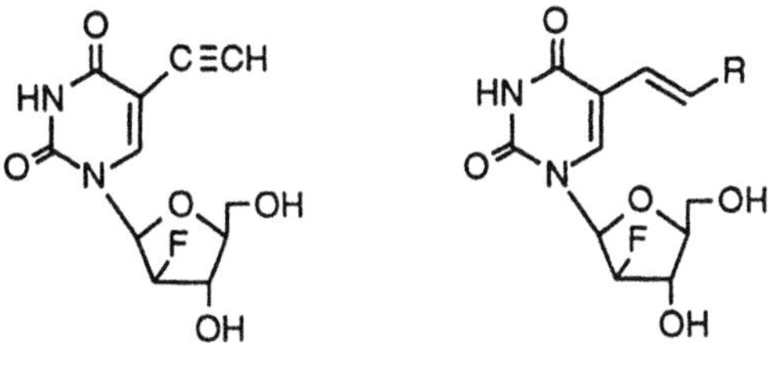

107

108: R = $COOCH_3$
109: R = COOH
110: R = Br

Figure 3.

using methanolic ammonia provided the targeted nucleoside analogs **95–106** (Scheme 13).

An acetylenic derivative was also prepared from the 3′,5′-di-*O*-benzoyl-2′-deoxy-2′-fluoro-5-iodo-β-L-arabinofuranosyluracil. Coupling of the 5-iodo analog **100** with trimethylsilylacetylene using $PdCl_2(PPh_3)_2$ and CuI as catalysts, followed by deprotection by sodium methoxide, led to the fully deprotected acetylenic derivative **107** (Figure 3). Starting from the 5-iodo analog **100**, coupling of methylacrylate using palladium diacetate $Pd(OAc)_2$ and PPh_3 gave the β-L-5-(*E*)-methoxycarbonylvinyl derivative **108**. Its saponification with NaOH, followed by acidification with HCl, gave the β-L-5-(*E*)-carboxyvinyl analog **109**, which is readily converted to the β-L-5-(*E*)-bromovinyl compound **110** using *N*-bromosuccinimide and K_2CO_3 (Figure 3).

B. 3′-Deoxy-L-nucleosides and Their Derivatives

Mathé et al.[37] stereospecifically synthesized the 3′-deoxy-β-L-*erythro*-pentofuranosyl nucleosides of the five naturally occurring nucleic bases **111–115** (Figure 4).

Starting from commercially available L-xylose, the 1,2-di-*O*-acetyl-3-deoxy-5-*O*-benzoyl-*erythro*-pentofuranose was prepared, and coupled to the pyrimidine and purine bases providing the 3′-deoxy-β-L-cytidine **111**, thymidine **112**, uridine **113**, adenosine **114**, and guanosine **115** (Figure 4).

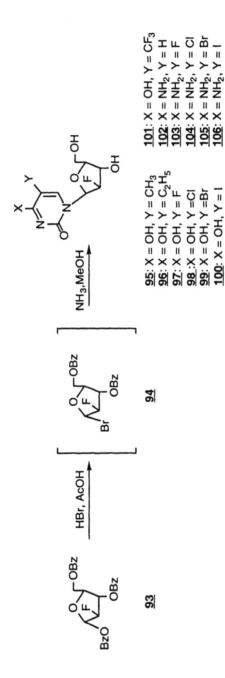

95: X = OH, Y = CH₃	**101**: X = OH, Y = CF₃
96: X = OH, Y = C₂H₅	**102**: X = NH₂, Y = H
97: X = OH, Y = F	**103**: X = NH₂, Y = F
98: X = OH, Y = Cl	**104**: X = NH₂, Y = Cl
99: X = OH, Y = Br	**105**: X = NH₂, Y = Br
100: X = OH, Y = I	**106**: X = NH₂, Y = I

Scheme 13.

22

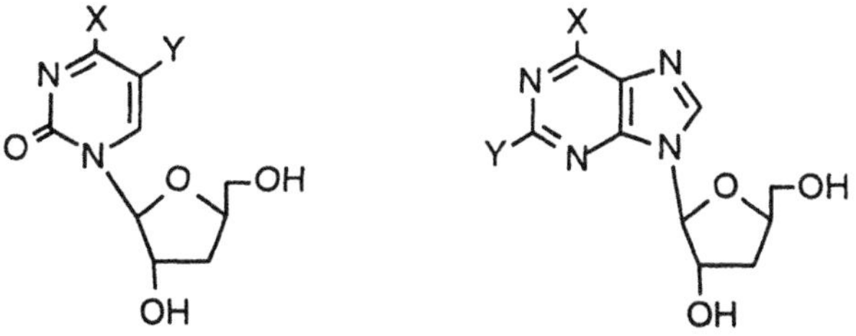

111: X = NH$_2$, Y = H
112: X = OH, Y = CH$_3$
113: X = OH, Y = H

114: X = NH$_2$, Y = H
115: X = OH, Y = NH$_2$

Figure 4.

Génu-Dellac et al.[18] prepared some 3′-deoxy-3′-substituted-α-L-lyxofuranosylpyrimidine nucleosides analogs (Scheme 14).

Starting from the 3′,5′-di-O-benzoyl-α-L-thymidine **23**, as previously illustrated in Scheme 5, this was accomplished by protecting the 2′-hydroxyl group as a monomethoxytetrahydropyrane. Deprotection of benzoyl groups using methanolic ammonia and then specific protection of the 5′-hydroxyl group using the monomethoxytrityl chloride in pyridine yielded derivative **116** (Scheme 14). Reaction of **116** with phenylthionochloride and then tributyltin hydride in the presence of AIBN led to 1-(3-deoxy-α-L-*threo*-pentofuranosyl)thymine **117**. Treatment of **116** by DAST in pyridine, followed by action of trifluoroacetic acid, provided the 1-(3-deoxy-3-fluoro-α-L-lyxofuranosyl)thymine **118** and the 1-(3-O-methyl-α-L-lyxofuranosyl)thymine **119** (Scheme 14). Finally, treatment of **116** by trifluoromethanesulfonate anhydride and then sodium azide afforded the 1-(3-deoxy-3-azido-α-L-lyxofuranosyl)thymine **120**. Reaction of **120** with triphenylphosphine and then NH$_4$OH led to the 1-(3-deoxy-3-amino-α-L-lyxofuranosyl)thymine **121** (Scheme 14).

IV. 2′,3′-DIDEOXY-L-NUCLEOSIDES

Since the 2′,3′-dideoxy-D-nucleosides have shown potent activities against various viruses such as HIV or HBV, several pyrimidine and purine 2′,3′-dideoxy-L-nucleoside analogs have been synthesized.

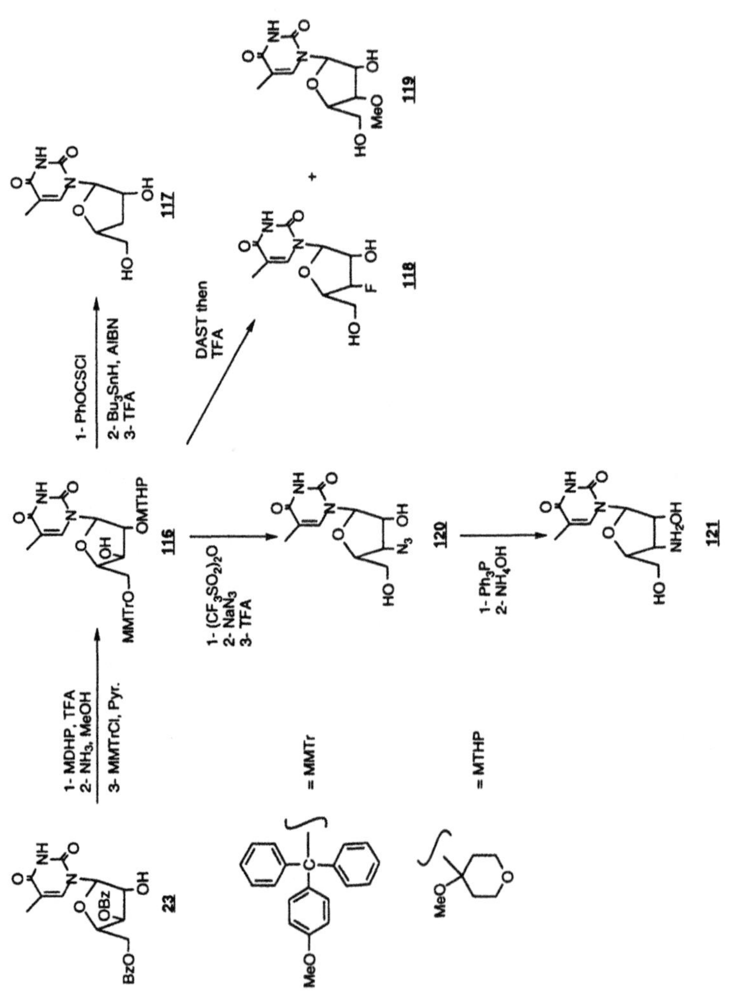

Scheme 14.

1- MDHP, TFA
2- NH₃, MeOH
3- MMTrCl, Pyr.

1- PhOCSCl
2- Bu₃SnH, AIBN
3- TFA

DAST then TFA

1- (CF₃SO₂)₂O
2- NaN₃
3- TFA

1- Ph₃P
2- NH₄OH

23

116

117

118

119

120

121

= MMTr

= MTHP

A. 2′,3′-Dideoxy-L-pyrimidine Nucleosides

Starting from enantiomerically pure 1-*O*-acetyl-2,3-dideoxy-L-ribofuranose **121**, Lin et al.[38,39] prepared a series of L-nucleoside analogs using regular or modified pyrimidine bases. The coupling step was achieved using a silylated base and various catalysts such as ethylaluminium chloride $AlEt_2Cl_2$ or potassium nonafluoro-1-butane sulfonate $C_4F_9SO_3K$. Subsequent deprotection using *n*-Bu_4NF in THF yielded to the α-L- and β-L-2′,3′-dideoxynucleosides **122–135** (Scheme 15).

Lin et al.[39] converted the 2′,3′-dideoxy-β-L-5-fluorouridine analog (**126**) into 2′,3′-dideoxy-β-L-5-fluorocytidine (**122**) by reacting the 5′-*O*-*t*-butyldimethylsilyl derivative of **126** in the presence of 1,2,4-triazole and 4-chlorophenylphosphorodichloridate in pyridine, followed by NH_4OH in dioxane, and final deprotection by *n*-Bu_4NF in THF (Scheme 15).

Stereospecific syntheses for some of the 2′,3′-dideoxy-L-nucleosides such as 2′,3′-dideoxy-β-L-cytidine **124** has been accomplished by Lin et al.[40] and Gosselin et al.[41] Starting from the 2,2′-anhydro-β-L-uridine **56**, described by Holý et al.,[20] Lin et al.[40] protected the 5′-hydroxyl group using *tert*-butyldimethylsilyl chloride in pyridine

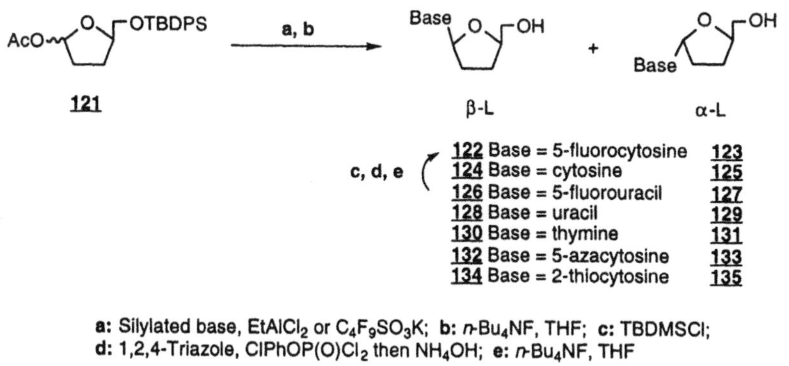
AcO
O
OTBDPS
121
a, b
Base
O
OH
+
O
OH
Base
β-L
α-L
c, d, e
122 Base = 5-fluorocytosine 123
124 Base = cytosine 125
126 Base = 5-fluorouracil 127
128 Base = uracil 129
130 Base = thymine 131
132 Base = 5-azacytosine 133
134 Base = 2-thiocytosine 135

a: Silylated base, $EtAlCl_2$ or $C_4F_9SO_3K$; **b:** *n*-Bu_4NF, THF; **c:** TBDMSCl; **d:** 1,2,4-Triazole, $ClPhOP(O)Cl_2$ then NH_4OH; **e:** *n*-Bu_4NF, THF

Scheme 15.

to yield compound **136**. Treatment of compound **136** with phenylchlo-
rothionocarbonate and then AIBN in presence of tributyltin hydride
(Bu$_3$SnH) provided the 5'-*O-t*-butyldimethylsilyl-2',3'-didehydro-
2',3'-dideoxy-β-L-uridine **137**. Hydrogenolysis with H$_2$ with Pd/C,
followed by treatment with 1,2,4-triazole and 4-chlorophenylphos-
phorodichloridate in pyridine, afforded, after final removal of silyl
protecting group, the desired 2',3'-dideoxy-β-L-cytidine **124** (Scheme
16).

Gosselin et al.[41] synthesized the 2',3'-dideoxy-β-L-cytidine **124** and
its 5-fluorocytidine derivative **122**, starting from the L-xylose which
allowed the preparation of both of these compounds.

Van Draanen et al.[42] prepared a series of 2',3'-dideoxy-α-L-
pyrimidine analogs **123**, **125**, **127**, **129**, **139–144** (Scheme 17). Using
the D-glutamic acid **138**, the same acetate **121** was obtained in five
steps. Nucleic acid base condensation was performed in presence of
BSA and TMSOTf and deprotection was realized using Et$_4$NF
(Scheme 17). Base condensation afforded α- and β-isomers which
were separated by preparative HPLC and column chromatography.

Van Draanen[42] used an enzymatic resolution with *E. coli* cytidine
deaminase (pH = 7.5) in order to overcome difficulties in separating
the α/β isomers of analogs **129** and **131**. Starting from the intermediate

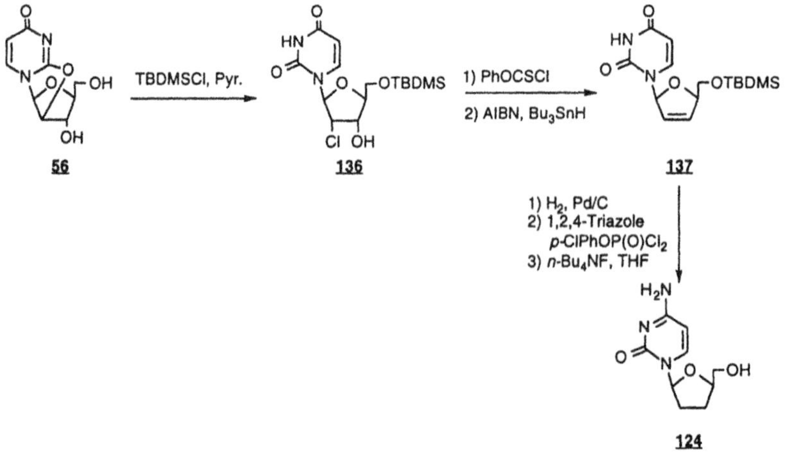

Scheme 16.

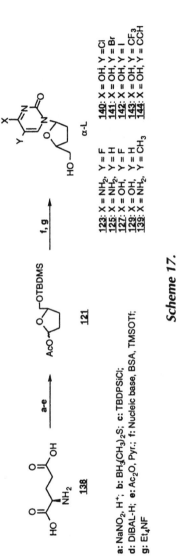

138 $\xrightarrow{\text{a-e}}$ **121** $\xrightarrow{\text{f, g}}$ α-L

123: X = NH₂, Y = F
125: X = NH₂, Y = H
127: X = OH, Y = F
129: X = OH, Y = H
139: X = NH₂, Y = CH₃

140: X = OH, Y = Cl
141: X = OH, Y = Br
142: X = OH, Y = I
143: X = OH, Y = CF₃
144: X = OH, Y = CCH

a: NaNO₂, H⁺; b: BH₃(CH₃)₂S; c: TBDPSiCl;
d: DiBAL-H; e: Ac₂O, Pyr.; f: Nucleic base, BSA, TMSOTf;
g: Et₄NF

Scheme 17.

27

121 condensed with cytosine and 5-methylcytosine, respectively, the
α/β-mixture of 2′,3′-dideoxy-L-cytidine **124/125** and 2′,3′-dideoxy-L-
5-methylcytidine **145/139** were synthesized. *E. coli* cytidine deaminase selectively deaminated the α-isomer of these mixtures much more
rapidly than the corresponding β-L-isomers. While the β-L-isomers remained unchanged as the cytosine derivatives (**124** and **145**), α-L-
isomers **125** and **139** were converted into their respective uracil
derivatives **129** and **131**, which were easily separated from the unreacted β-cytidine analogs (**124** and **145**, Scheme 18).

Finally, Mansuri et al.[43] also reported the synthesis of an α/β-L-
2′,3′-dideoxycytidine nucleosides (α/β-L-ddC) using a similar approach to that described by Van Draanen et al.[42] Mansuri et al. used as
intermediate 1-bromo-2,3-dideoxy-L-ribofuranose and the same synthetic route as described on Scheme 17.

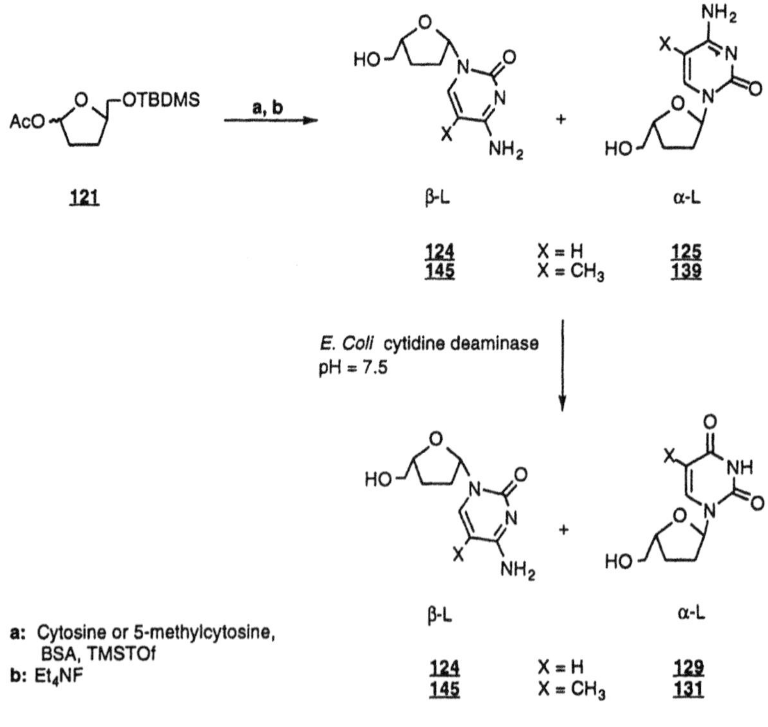

Scheme 18.

B. 2′,3′-Dideoxy-L-purine Nucleosides

Lin et al.[44] prepared a series of α- and β-2′,3′-dideoxy-L-purine nucleosides such as 2′,3′-dideoxy-L-adenosine (**156** and **157**), 2′,3′-dideoxy-2-chloro-L-adenosine (**158** and **159**), 1-(2,3-dideoxy-L-ribofuranosyl)-2-amino-6-chloropurine (**160** and **161**), 2′,3′-dideoxy-2-amino-L-adenosine (**162** and **163**) (Scheme 19), and N^7/N^9 isomers of 2′,3′-dideoxy-L-guanosine (**168-171**) (Scheme 20).

Starting form the 1-*O*-acetyl-5-*O*-*tert*-butyldimethylsilyl-2′,3′-dideoxyribofuranose **149**, Lin et al.[44] condensed various purine bases (**146–148**) using NaH and then $AlEt_2Cl_2$, followed by separation of the isomers. The corresponding 2′,3′-dideoxy-L-purine nucleosides were then obtained as α- and β-isomers (**150–155**, Scheme 19). Reaction of **150** (α-isomer) and **151** (β-isomer) with *n*-Bu₄NF in THF provided the deprotected 1-(2,3-dideoxy-L-ribofuranosyl)-2-amino-6-chloropurine **160** and **161** (α- and β-isomers), respectively. Treatment of **152** (α-isomer), **153** (β-isomer), **154** (α-isomer) and **155** (β-isomer) successively with methanolic ammonia and then *n*-Bu₄NF in THF yielded the corresponding isomers of 2′,3′-dideoxy-L-adenosine **156–157** and 2′,3′-dideoxy-2-chloro-L-adenosine **158–159** (Scheme 19). Reaction of **150** and **151** or **154** and **155** with first lithium azide, then reduction using $LiAlH_4$, and removal of the protecting silyl group afforded the corresponding α- and β-isomers of 2′,3′-dideoxy-2-amino-L-adenosine (**162** and **163**, Scheme 19).

Using the same intermediate 1-*O*-acetyl-5-*O*-*tert*-butyldimethyl-silyl-2′,3′-dideoxyribofuranose **149** and TMSOTf, Lin et al.[44] condensed the silylated guanine and obtained α- and β-isomers of the N^7 and N^9 of 2′,3′-dideoxy-L-guanosine (**168–171**, Scheme 20).

Treated independently with *n*-Bu₄NF in THF and then NH_3 in MeOH, the four isomers **164–167** yielded to the corresponding deprotected β-N^9 (**168**), α-N^9 (**169**), β-N^7 (**170**), and α-N^7 (**171**) isomers of 2′,3′-dideoxy-L-guanosine (Scheme 20).

Bolon et al.[45] also reported the synthesis of the 2′,3′-dideoxy-β-L-adenosine (**157**), 2′,3′-dideoxy-β-L-guanosine (**168**) and the 2′,3′-dideoxy-β-L-inosine (**176**) (Scheme 21). They used a convenient route allowing to synthesize the 2′,3′-dideoxy- and the 2′,3′-didehydro-2′,3′-dideoxy-L-nucleosides, starting from the D-glutamic acid or the L-xylose as β-isomer in very high stereoselectivity. The key reaction

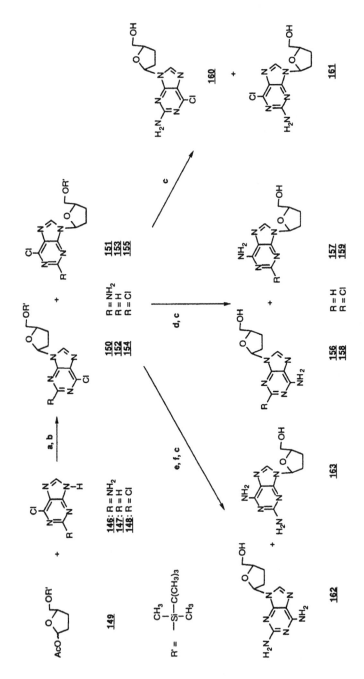

a: NaH; b: AlEt₂Cl₂; c: n-Bu₄NF, THF; d: NH₃, MeOH; e: LiN₃, EtOH; f: LiAlH₄

Scheme 19.

30

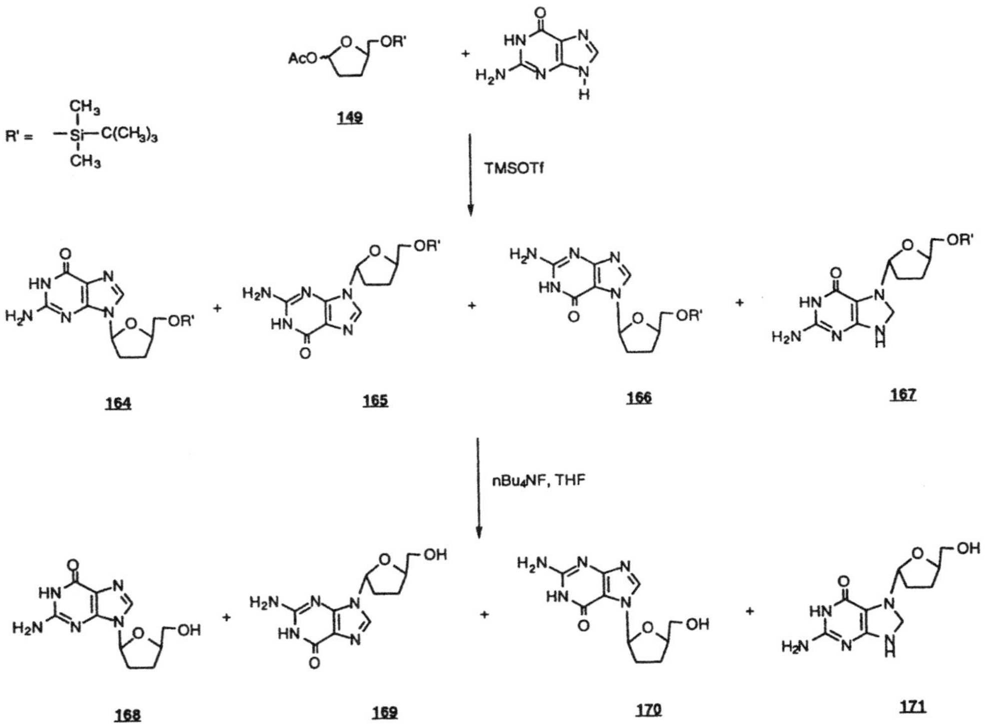

Scheme 20.

31

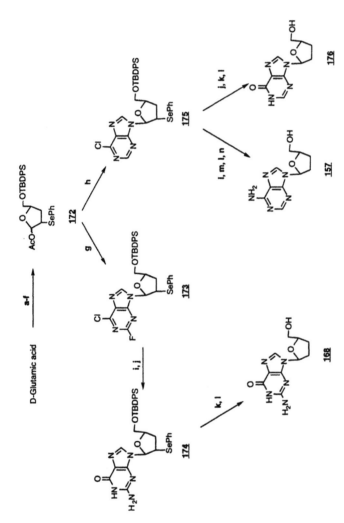

a: NaNO₂, HCl; b: BH₃, THF; c: TBDPSiCl, imidazole; d: 1-LiHMDS, -78°C; 2- TMSCl; 3-PhSeBr; e: DIBAL-H f: Ac₂O, pyridine;
g:TMS-2-fluoro-6-chloropurine, TMSOTf; h: TMS-6-chloropurine, TMSOTf; j: HS-(CH₂)₂-OH, NaOMe, MeOH; k: Bu₃SnH, Et₃B
l: n-Bu₄NF, THF; m: H₂O₂, Pyr.; n: H₂, Pd/C

Scheme 21.

32

of this synthesis was the introduction of a phenylselenenyl group or the functionalization of the 2'-hydroxy group by phenylthionocarbonate chloride or mesyl chloride. Reported here is only one synthetic pathway, since other routes led to the same desired 2',3'-dideoxy-β-L-purine nucleoside analogs **157**, **168**, and **176**. Reaction of the 2'-phenylselenenyl derivative, which was achieved using tributyltin hydride with tributyl borane, provided the desired 2',3'-dideoxyfuranosyl moiety, while its treatment with H_2O_2 in pyridine led to the 2',3'-didehydro-2',3'-dideoxyfuranosyl ring as described by Beach et al.[46] D-Glutamic acid was converted in three steps into the 5-*O*-*t*-butyldiphenylsilyl-2,3-dideoxy-pentofuranone. Phenylselenenyl group was introduced at position 2 using LiHMDS and then phenylselenenyl bromide at −78 °C. Further reduction with DIBAL-H and then acetylation of the lactol provided the 1-*O*-acetyl-2-phenylselenenyl-5-*O*-*tert*-butyldiphenylsilyl-2,3-dideoxy-pentofuranose **172** (Scheme 21).

Condensation with silylated 2-fluoro-6-chloropurine using TMSOTf led to the nucleoside analog **173**. Successive treatments of **173** with methanolic ammonia and 2-thioethanol with sodium methoxide in methanol led to the protected 2',3'-dideoxyguanosine intermediate **174**. Reaction of **174** with tributyltin hydride in presence of tributylborane and, finally, with *n*-Bu$_4$NF in THF, yielded the desired 2',3'-dideoxy-β-L-guanosine **168** (Scheme 21). Condensation of **172** with silylated 6-chloropurine in presence of TMSOTf as catalyst produced the 1-(2-phenylselenenyl-5-*O*-*tert*-butyldiphenylsilyl-2,3-dideoxy-β-L-pentofuranosyl)-6-chloropurine **175**. Following treatments with 2-thioethanol in presence of sodium methoxide in methanol, tributyltin hydride in presence of tributylborane, and then with *n*-Bu$_4$NF in THF, afforded the 2',3'-dideoxy-β-L-inosine **176** (Scheme 21). Reaction of **175** with NH_3/MeOH, followed by H_2O_2 in pyridine, the removal of the silyl protecting group, and then catalytic hydrogenation on Pd/C, led to the desired 2',3'-dideoxy-β-L-adenosine **157** (Scheme 21).

Mansuri et al.[43] also synthesized the α- and β-L-ddA (**156** and **157**) using the methodology used later by Van Draanen et al.[42] for the synthesis of α- and β-L-ddC.

El Alaoui et al.[47] reported the synthesis of a series of β-L-2',3'-dideoxypurine analogs **157**, **162**, **168**, **176–180** (Figure 5).

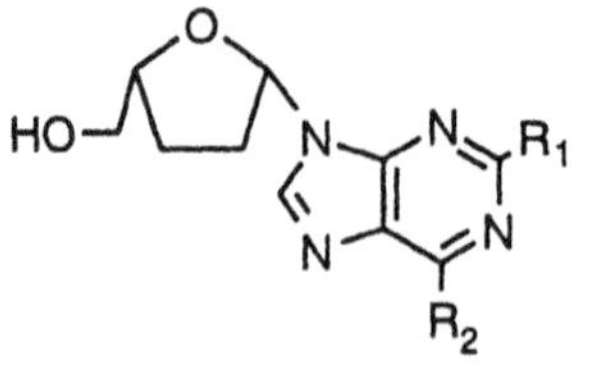

157: R_1 = H, R_2 = NH$_2$ **177**: R_1 = H, R_2 = cyclopropylamino-
162: R_1 = NH$_2$, R_2 = NH$_2$ **178**: R_1 = NH$_2$, R_2 = methylamino-
168: R_1 = NH$_2$, R_2 = OH   **179**: R_1 = NH$_2$, R_2 = cyclopropylamino-
176: R_1 = H, R_2 = OH   **180**: R_1 = NH$_2$, R_2 = cyclopentylamino-

Figure 5.

Van Draanen et al.[42] reported the synthesis of 2′,3′-dideoxy-α-L-purine nucleosides using thymidine phosphorylase (TPase) and PNP-catalyzed transfer of sugars from pyrimidine nucleosides to purine nucleosides as described in previous studies.[48] The enzyme transfer resulted in total retention of the α-L configuration of the dideoxypentosyl moiety from the α-L-dideoxyuridine **129** to various purine bases (**156, 162, 181–182**, Scheme 22). The transfer depended on the concentration of enzyme and was achieved in 60 to 80% yield.

α-L-Hypoxanthine **183** and α-L-guanine **169** derivatives were obtained by treatment of compound **156** and **162** by calf intestinal adenosine deaminase (Scheme 23).

Lee et al.[49] prepared some 2′,3′-dideoxy-L-nucleosides wherein the nucleic base was replaced by a modified pyrimidine or purine base,

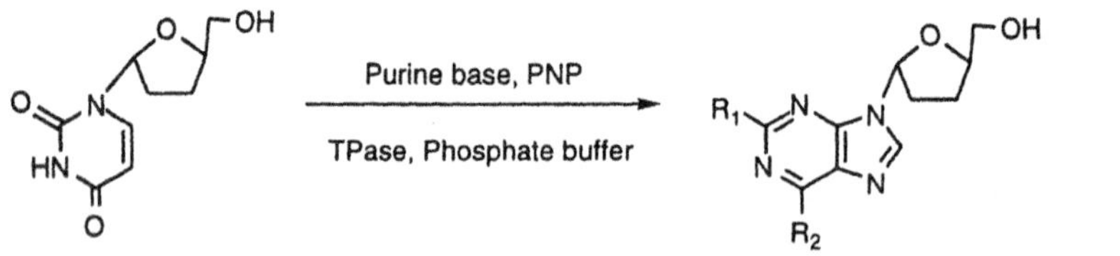

Purine base, PNP
⟶
TPase, Phosphate buffer

129

156: R_1 = H, R_2 = NH$_2$
162: R_1 = NH$_2$, R_2 = NH$_2$
181: R_1 = H, R_2 = OCH$_3$
182: R_1 = NH$_2$, R_2 = OCH$_3$

Scheme 22.

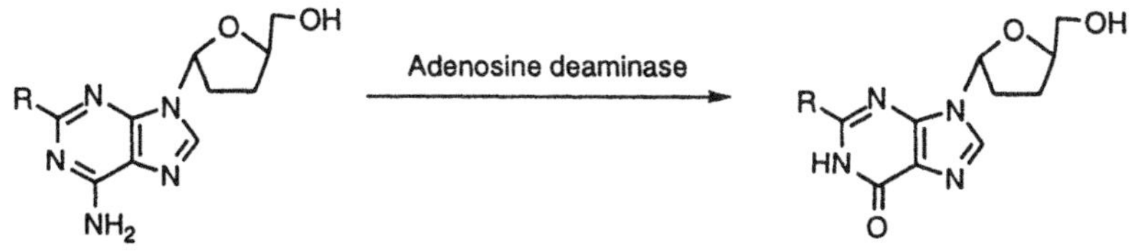

Scheme 23.

such as 5-aza-9-deazaadenine, leading to novel "*C*-nucleosides" **188**, **189**, **193**, **194**, **196**, and **197**. The synthetic route chosen was to build the modified base directly on the pseudo sugar moiety. They converted the L-gulonic-γ-lactone into a 5-hydroxymethyl-2,3-dideoxy-γ-lactone **184** in five steps.[46] Protection of the hydroxyl group using trityl chloride in pyridine and then DIBAL-H yielded the 5-*O*-trityl-2,3-dideoxy-L-γ-lactol **185** (Scheme 24). Treatment of derivative **185** with sodium hydride and diethylcyanomethylphosphonate, followed by reaction with bis(dimethylamino)-*tert*-butoxymethane, yielded compound **186**. When reacted in presence of hydrazine, **186** cyclized to the 3-aminopyrazole derivative **187**, which afforded, after treatment with methyl-*N*-cyanomethanimidate and then HCl in methanol, the desired 5-aza-9-deaza-adenine *C*-nucleosides **188** (β) and **189** (α) (Scheme 24).

Compound **186** was converted to its hydroxyl derivative **190** by treatment with trifluoroacetic acid. Successive treatments of **190** with aminoacetonitrile and sodium acetate, DBN and ethylchloroformate, DBN, and, finally, with sodium carbonate in methanol, provided the pyrole derivative **191** (Scheme 25). Reaction of **191** with formamidine in boiling MeOH led to the 9-deaza-adenine derivative **192** which, after deprotection, yielded the 9-deaza-adenine *C*-nucleosides **193** (β) and **194** (α) (Scheme 25).

Hypoxanthine analogs were obtained starting from **190**. Treatment with ethyl glycinate with sodium acetate, followed by methylchloroformate, sodium ethoxide, and then reaction of **190** with formamidine in boiling ethanol, produced to the pyrrolo[3,2-d]pyrimidine derivative **195** as α/β mixture (Scheme 25). Further treatment of **195** with

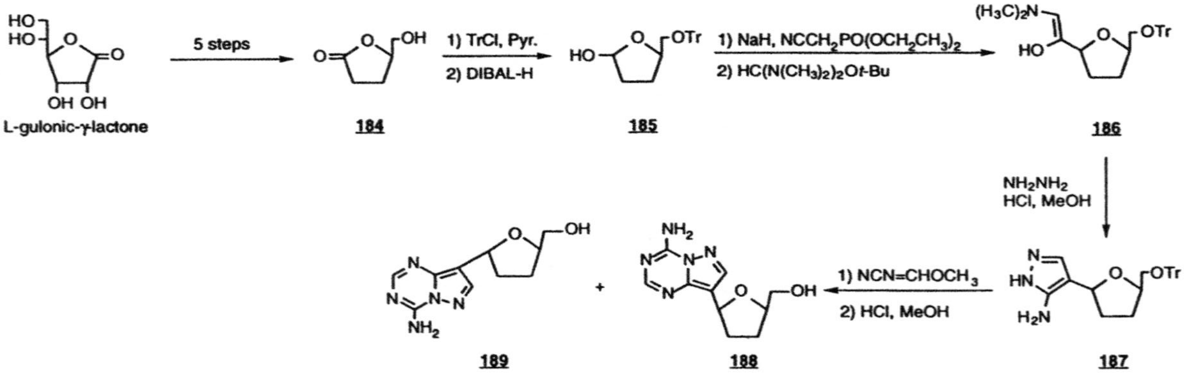

L-gulonic-γ-lactone

5 steps

184

1) TrCl, Pyr.
2) DIBAL-H

185

1) NaH, NCCH$_2$PO(OCH$_2$CH$_3$)$_2$
2) HC(N(CH$_3$)$_2$)$_2$Ot-Bu

186

NH$_2$NH$_2$
HCl, MeOH

187

1) NCN=CHOCH$_3$
2) HCl, MeOH

188

+

189

Scheme 24.

36

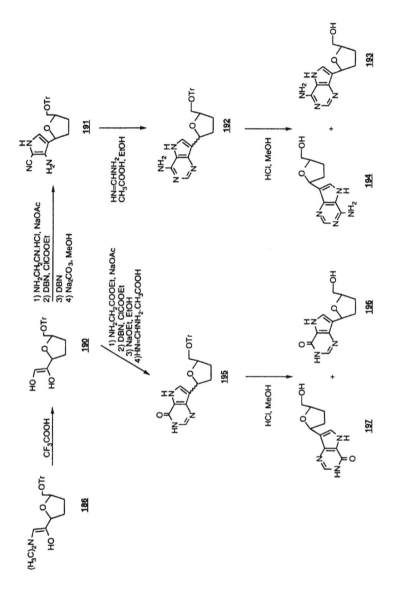

Scheme 25.

37

HCl in methanol provided the 9-deaza hypoxanthine *C*-nucleosides
196 (β) and **197** (α) (Scheme 25).

C. 2′ and/or 3′-Substituted-2′,3′-dideoxy-ʟ-nucleosides

Xiang et al.[50] synthesized the 1- or 9-(2,3-dideoxy-2-fluoro-β-ʟ-
threo-pentofuranosyl)cytosine or adenine derivatives in which the
2-position of the sugar moiety was substituted by a fluorine (Scheme
26). Starting from the ʟ-xylose, Xiang et al. synthesized the 1,2 and
3,5-isopropylidene **45**, as described by Gosselin et al.,[22] and then
specifically deprotected the 3,5-acetonide and reprotected the
5-hydroxymethyl with a benzoyl group, leaving the 3-hydroxyl
group free (**198**, Scheme 26). Oxidation of the 3-hydroxy of **198** with
pyridinium dichromate in acetic anhydride led to the 3-oxo derivative
199. Successive treatment with tosylhydrazine, sodium cyanoborohy-
dride, and then sodium acetate in ethanol afforded the 3-deoxy derivative **200** (Scheme 26). Compound **200** was then converted to the
1-bromo-2-fluoro-5-benzoyloxymethyl-*threo*-pentofuranose **201** using acetic acid, DAST with DMAP and then hydrobromic acid. This
key intermediate was condensed with silylated cytosine in the presence of TMSOTf. Removal of the benzoyl group by NH_3/MeOH
yielded 1-(2,3-dideoxy-2-fluoro-β-ʟ-*threo*-pentofuranosyl)cytosine
202. When condensed with 6-chloropurine compound using NaH,
followed by deprotection by NH_3/MeOH at 100 °C, **201** yielded to the
9-(2,3-dideoxy-2-fluoro-β-ʟ-*threo*-pentofuranosyl)-adenine **203**
(Scheme 26).

Génu-Dellac et al.[18] also reported the synthesis of some 3′-substi-
tuted-2′,3′-dideoxy-ʟ-nucleosides. Starting from the 2′-deoxy-5′-*O*-
monomethoxytrityloxy-α-ʟ-thymidine **204** (Scheme 27),
2′,3′-dideoxy-3′-azido-α-ʟ-lyxothymidine **205** was prepared using
successively trifluoromethanesulfonate anhydride, sodium azide in
DMF, and then trifluoroacetic acid. When treated with triphenylphosphine in pyridine and then ammonium hydroxide, compound **205**
led to the 2′,3′-dideoxy-3′-amino-α-ʟ-lyxothymidine **206** (Scheme
27). Treatment with DAST in pyridine and then trifluoroacetic acid,
204 afforded the 2′,3′-dideoxy-3′-fluoro-α-ʟ-lyxothymidine **207**.

Wengel et al.[51] synthesized the four stereoisomers of ʟ-3′-azido-3′-
deoxythymidine (Scheme 28). ʟ-Arabinose was first transformed into

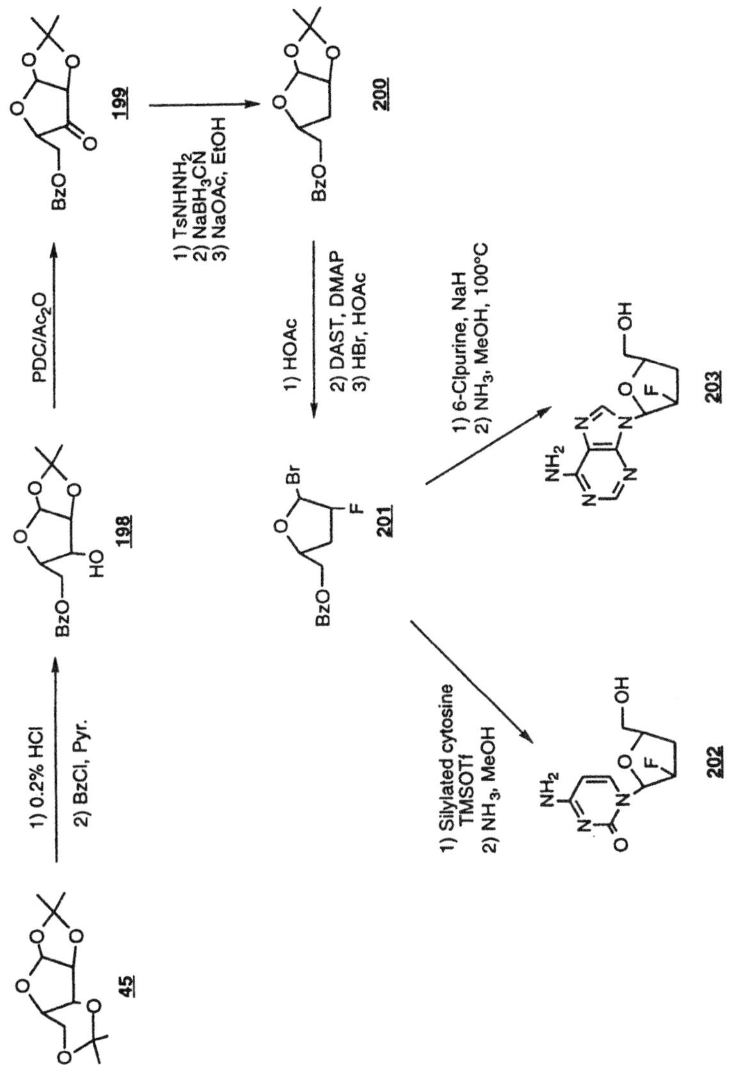

45
1) 0.2% HCl
2) BzCl, Pyr.
BzO
HO
198
PDC/Ac2O
BzO
199
1) TsNHNH2
2) NaBH3CN
3) NaOAc, EtOH
BzO
200
1) HOAc
2) DAST, DMAP
3) HBr, HOAc
BzO
Br
F
201
1) Silylated cytosine
TMSOTf
2) NH3, MeOH
NH2
OH
F
202
1) 6-Clpurine, NaH
2) NH3, MeOH, 100°C
NH2
OH
F
203

Scheme 26.

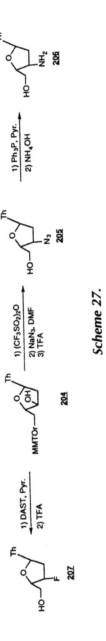

Scheme 27.

40

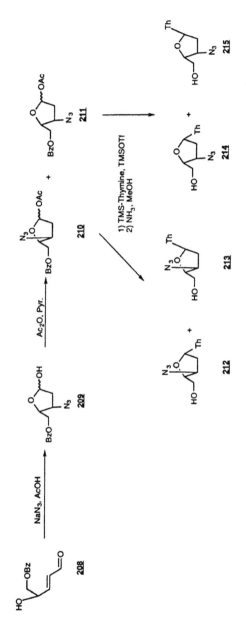

Scheme 28.

41

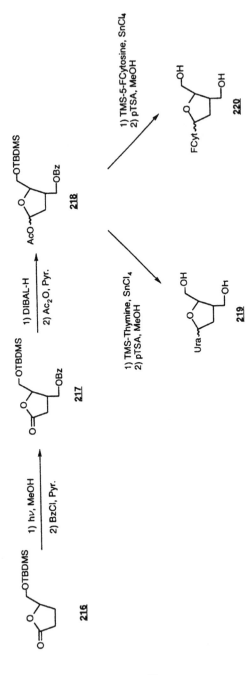

Scheme 29.

42

4-hydroxy-5-*O*-benzoyl-pent-2,3-enone **208**, which was converted to the 1-hydroxy-3-azido-5-*O*-benzoyl-2,3-dideoxy-L-*erythro*- and -threo-pentofuranose **209**. Direct acetylation of the lactol with acetic anhydride and DMAP in pyridine afforded, after purification on column, the 1-*O*-acetyl-3-azido-5-*O*-benzoyl-2,3-dideoxy-L-*erythro*-pentofuranose **210** and 1-*O*-acetyl-3-azido-5-*O*-benzoyl-2,3-dideoxy-L-*threo*-pentofuranose **211** (Scheme 28). Condensation of **210** and **211** with silylated thymine in presence of TMSOTf as catalyst, followed by removal of the benzoyl group using NH_3/MeOH, provided, respectively, the β-L-*erythro* (**212**), α-L-*erythro* (**213**), β-L-*threo* (**214**), and α-L-*threo* (**215**) derivatives of 3'-azido-3'-deoxythymidine (AZT) (Scheme 28).

Using different synthetic routes, Czernecki et al.[52] reported the synthesis of 3'-azido-3'-deoxy-*threo*-L-thymidine **214**, while Sugimura et al.[53] reported routes for the preparation of the *erythro* derivatives **212** and **213**.

Gould et al.[54] synthesized 3'-hydroxymethyl-2',3'-dideoxy-L-nucleoside analogs with uracil and 5-fluorocytosine bases as an α/β mixture. When irradiated with light, the 5(*R*)-*tert*-butyldimethyl-siloxymethyl-2(5*H*)-furan-2-one **216** in MeOH produced the 3-benzoyloxymethyl-5-*O*-*tert*-butyldimethylsilyl-2,3-dideoxy-furan-2-one **217**. Reduction with DiBAL-H and then acetylation of the lactol afforded the 1-*O*-acetyl-3-benzoyloxymethyl-5-*O*-*tert*-butyldimethylsilyl-2,3-dideoxy-furan **218** (Scheme 29). Condensation of **218** with

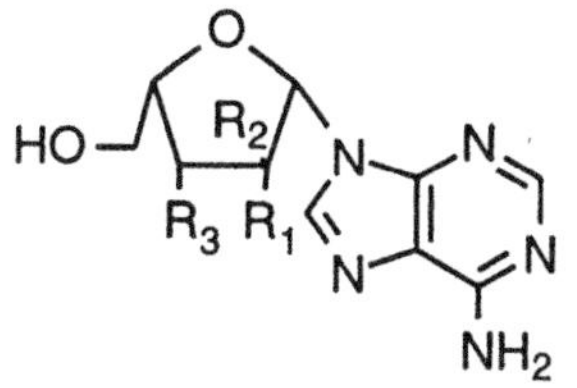

221: R_1 = H, R_2 = N_3, R_3 = H   **225**: R_1 = F, R_2 = H, R_3 = H
222: R_1 = H, R_2 = NH_2, R_3 = H   **226**: R_1 = H, R_2 = H, R_3 = N_3
223: R_1 = N_3, R_2 = H, R_3 = H   **227**: R_1 = H, R_2 = H, R_3 = NH_2
224: R_1 = NH_2, R_2 = H, R_3 = H   **228**: R_1 = H, R_2 = H, R_3 = F

Figure 6.

silylated uracil or silylated 5-fluorocytosine in presence of tin(IV) chloride as catalyst, followed by the deprotection with *p*-toluene sulfonamide in MeOH, led to the L-uracil analog **219** and 5-L-fluorocytosine derivative **220** as an α/β mixture.

El Alaoui et al.[47] also synthesized a series of 2' or 3'-substituted-2',3'-dideoxy-β-L-adenine nucleosides with fluorine, azide, or amine as substituents (**221–228**, Figure 6).

V. 2',3'-UNSATURATED-L-NUCLEOSIDES

A. 2',3'-Didehydro-2',3'-dideoxy-L-nucleosides

Starting from the 3',5'-dibenzoyl-2-deoxy-L-uridine **71**, Lin et al.[55] reported the synthesis of cytosine derivatives such as β-L-d4C (**234**) and β-L-d4FC (**235**). Nucleoside **71** was also converted to the 5-fluoro derivative **229** after treatment of **71** with silylated 5-fluorouracil in presence of TMSOTf (Scheme 30). After reaction of **71** or **229** by methanolic ammonia, followed by mesyl chloride and then 1N NaOH in ethanol–water, the uracil **230** and 5-fluorouracil **231** derivatives of 3',5'-*O*-anhydronucleoside were obtained, respectively. Treatment with triazole in presence of *p*-chlorophenylphosphodichloridate and then ammonium hydroxide yielded the 3',5'-*O*-anhydro nucleoside derivative of cytosine **232** and 5-fluorocytosine **233** (Scheme 30). Final opening of 3',5'-anhydride with potassium *tert*-butylate in DMSO provided the final d4N nucleosides **234** (d4C) and **235** (d4FC).

Mansuri et al.[43] also synthesized some pyrimidine d4-nucleoside analogs. Coupling of the L-isomer of **10**—already used by Cusack et al.[14] (Scheme 2) with silylated thymine or uracil under Vorbrüggen conditions—followed by treatment by methanolic ammonia and then acetoxy-*iso*-butyryl bromide, led to the bromoacetoxy derivative **236** and **237**, respectively (Scheme 31). Reaction of **236** or **237** with Zn/Cu in DMF, followed by treatment with methanolic ammonia, yielded the corresponding d4-nucleosides **238** and **239**. Treatment of **239** with benzoyl chloride, followed by treatment with Lawesson's reagent in CHCl$_3$ at reflux, led to the thioamide derivative **240**. One-pot deprotection and amination at the 4-position of **240** in methanolic ammonia provided d4C (**234**, Scheme 31).

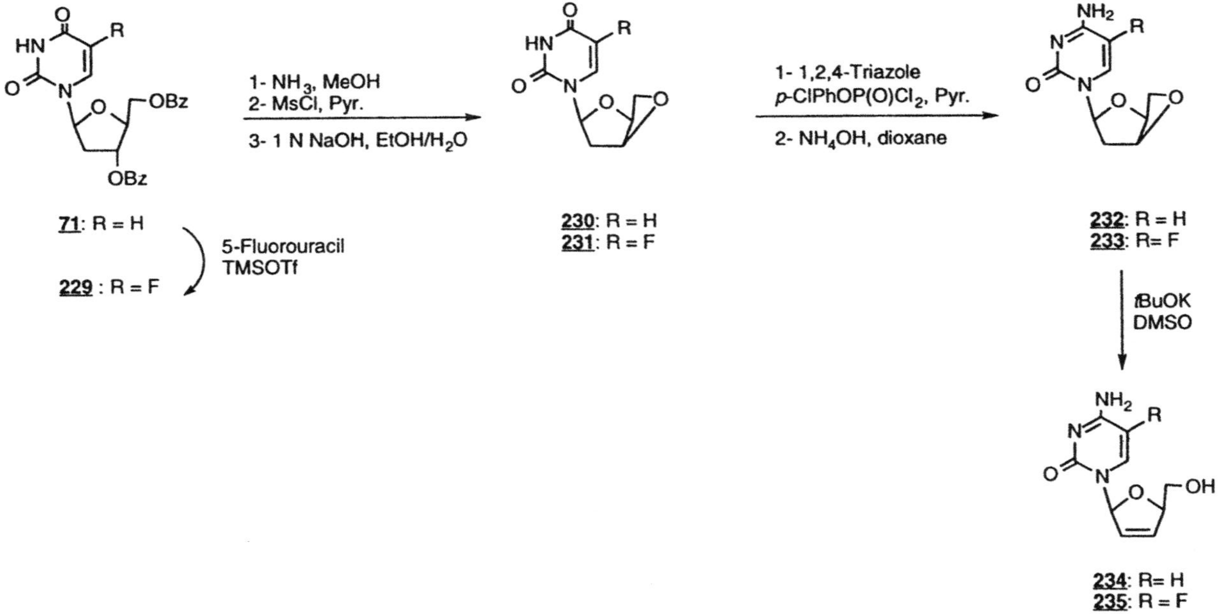

1- NH₃, MeOH
2- MsCl, Pyr.
3- 1 N NaOH, EtOH/H₂O

71: R = H

229 : R = F

5-Fluorouracil
TMSOTf

230: R = H
231: R = F

1- 1,2,4-Triazole
p-ClPhOP(O)Cl₂, Pyr.

2- NH₄OH, dioxane

232: R = H
233: R = F

*t*BuOK
DMSO

234: R = H
235: R = F

Scheme 30.

45

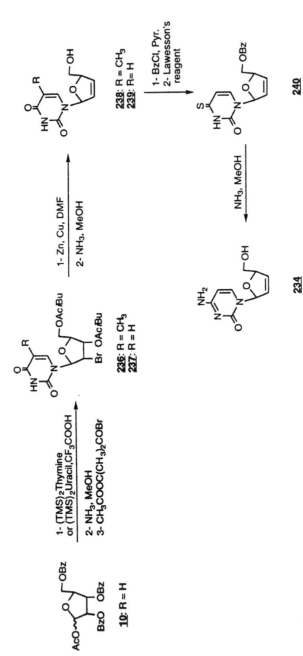

Scheme 31.

46

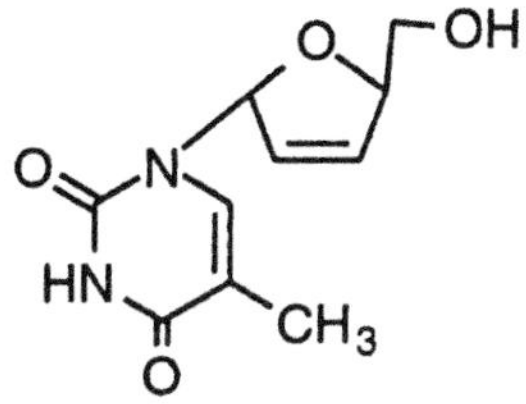

Figure 7.

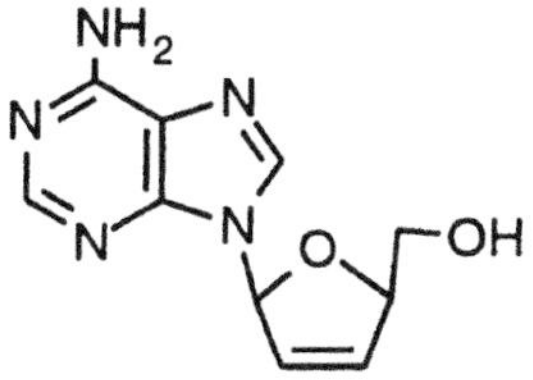

Génu-Dellac et al.[18] and El Alaoui et al.,[47] respectively, reported the synthesis of the α-L-d4T (**25**) and the β-L-d4A (**241**, Figure 7).

Bolon et al.[45] synthesized some purine d4-nucleosides such as d4A (**241**, Scheme 32), d4G (**242**), and d4I (**243**, Scheme 33). Converting the L-xylose into the 1,2-*O*-acetyl-5-*O*-benzoyl-L-pentofuranose **244** in five steps, the adenine derivative **245** was obtained after coupling **244** with silylated adenine in presence of tin(IV) chloride. Successive treatment of **245** with sodium methoxide and mesyl chloride led to the 2'-mesyl derivative **246**. Reacting with TBAF in THF and, finally, with methanolic ammonia, compound **246** yielded to the β-L-2',3'-didehydro-2',3'-dideoxyadenosine **241** (Scheme 32).

Coupling of 6-chloropurine or 2-fluoro-6-chloropurine with the intermediate **172** in presence of TMSOTf provided the intermediates **173** and **175**, respectively (Schemes 21 and 33). Treatment with methanolic ammonia and then mercaptoethanol/sodium methoxide, followed by reaction with H_2O_2 and final removal of the protecting group with TBAF in THF, provided the β-L-2',3'-didehydro-2',3'-dideoxyguanosine **242** (d4G). Treatment of **175** with mercaptoethanol/sodium methoxide, followed by hydrogen peroxide and TBAF in THF, yielded the β-L-2',3'-didehydro-2',3'-dideoxyinosine **243** (d4I) (Scheme 33).

B. Modified 2',3'-didehydro-2',3'-dideoxynucleosides

Graciet et al.[56] recently reported the coupling of **172** (see Scheme 33), with 5-carboranyluracil[57,58] in presence of TMSOTf or tin(IV) chloride as catalyst, yielding the α-L- (**247**) and β-L-5-*o*-carboranyl-

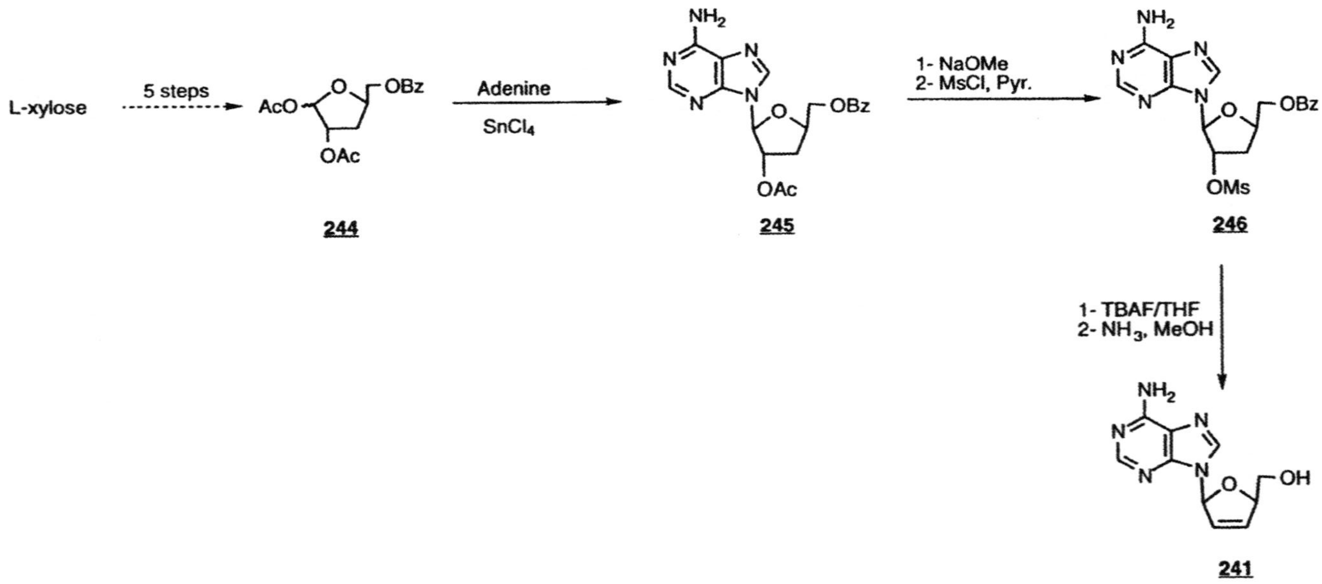

Scheme 32.

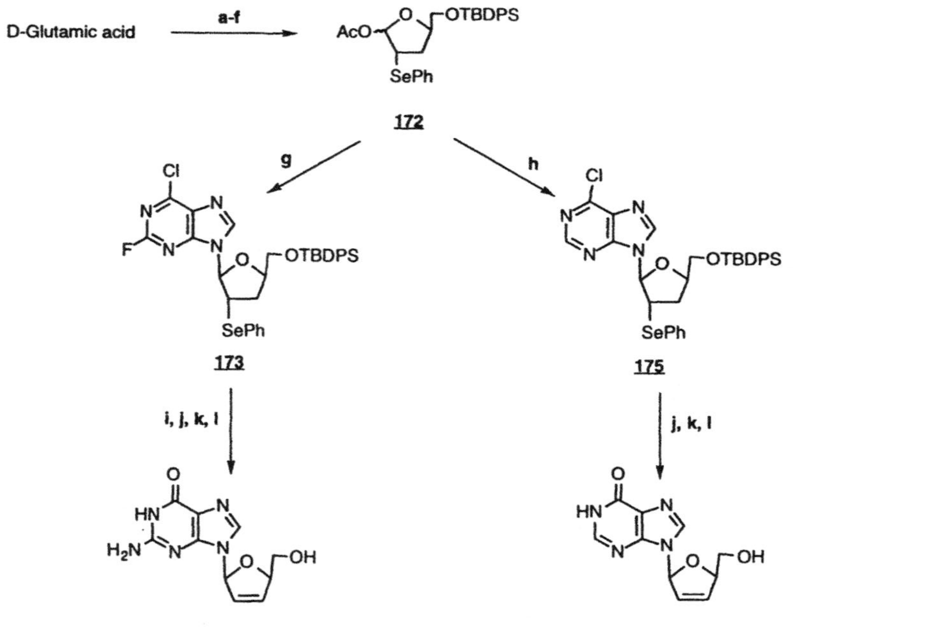

Scheme 33.

a: NaNO₂, HCl; b: BH₃, THF; c: TBDPSiCl, imidazole; d: 1-LiHMDS, -78°C; 2- TMSCl; 3-PhSeBr; e: DIBAL-H f: Ac₂O, pyridine; g: TMS-2-fluoro-6-chloropurine, TMSOTf; h: TMS-6-chloropurine, TMSOTf; i: NH₃, MeOH; j: HS-(CH₂)₂-OH, NaOMe, MeOH; k: H₂O₂, pyridine; l: Bu₄NF, THF.

5'-*O-tert*-butyldiphenylsilyl-2'-phenyl selenenyl-2',3'-didehydro-2',3'-dide oxy- uridine (**248**). After reaction with hydrogen peroxide in pyridine and treatment by *n*-Bu$_4$NF in THF, α-L- (**249**) and β-L-5-*o*-carboranyl-2',3'-didehydro-2',3'-dideoxyuridine (**250**) were obtained (Scheme 34).

VI. L-HETERONUCLEOSIDES

The nucleosides wherein the 4'-oxygen is replaced by other heteroatoms such as sulfur, for example, or the sugar moiety incorporates two heteroatoms, are not related to the natural nucleosides and their derivatives (D-) or their mirror images (L-). For convenience, these heteronucleosides (4'-thionucleosides or 1,3-oxathiolanyl derivatives) are called "D-" or "L-" relatively to the configuration of the 4'-carbon, because of the similar configuration seen in D- or L-nucleosides, respectively. In the D-heteronucleosides, this 4'-carbon is in an *S* configuration, while in the L-heteronucleosides, this carbon is in an *R* configuration. In the case of the 1,3-dioxolanyl analogs, the L-nucleosides are in a C4' *S* configuration and the D-nucleosides in a C4' *R* configuration.

A. One Heteroatom Containing Nucleosides

Since 4'-thio-D-nucleosides have demonstrated interesting biological properties,[59] Lin et al.[44] synthesized a series of various 2',3'-dideoxy-4'-thio-L-purine nucleosides. Starting from a 1-*O*-acetyl-5-*O-tert*-butyldimethylsilyl-2,3-dideoxy-4-thio-L-ribofuranose **251**, the synthesis of the desired nucleosides was achieved using a sodium salt of various purine bases in dry acetonitrile in the presence of ethyl aluminum dichloride or diethylaluminum chloride as catalyst, and provided the corresponding α/β nucleoside analogs **252/253** and **254/255** (Scheme 35). The following steps yielded various purine 4'-thionucleosides, depending on the nature of the used reagents (Scheme 35). Silyl group removal of **252** and **253** by TBAF in THF provided the deprotected 6-chloropurine analogs **256** and **257**. Further reaction of **256** and **257** with methanolic ammonia led to the adenine analogs **258** and **259**. Reaction of **254** and **255** with lithium azide in ethanol, followed by LiAlH$_4$ reduction, yielded, after final deprotec-

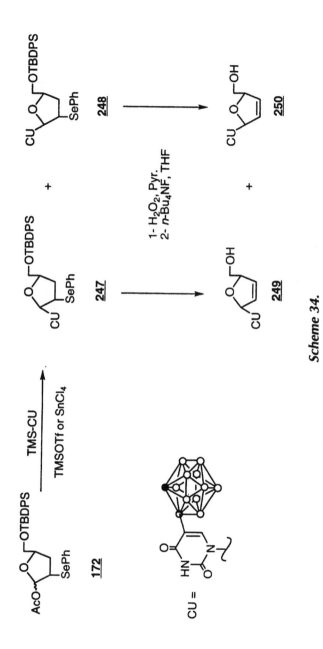

Scheme 34.

51

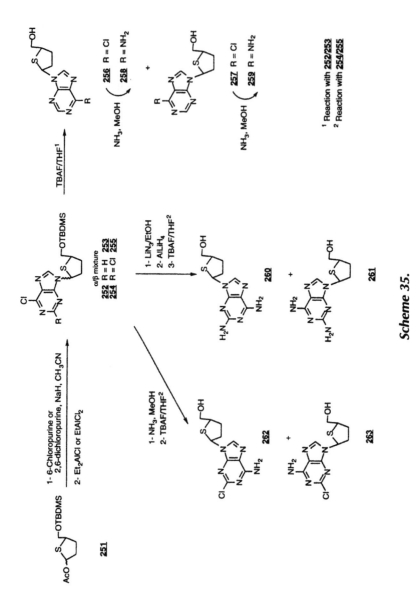

Scheme 35.

52

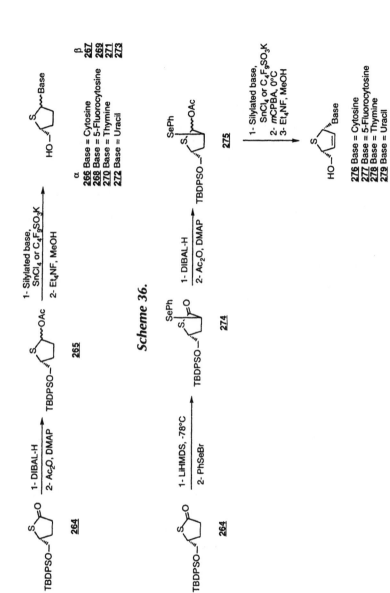

Scheme 36.

α
266 Base = Cytosine
268 Base = 5-Fluorocytosine
270 Base = Thymine
272 Base = Uracil

β
267
269
271
273

276 Base = Cytosine
277 Base = 5-Fluorocytosine
278 Base = Thymine
279 Base = Uracil

Scheme 37.

tion with TBAF/THF, the 2,6-diaminopurine-4'-thionucleosides **260** and **261**. Treatment of **254** and **255** with methanolic ammonia and then TBAF/THF provided the 2-chloroadenine derivatives **262** and **263** (Scheme 35).

Starting from the enantiomerically pure thiolactone **264** which was obtained in two steps from the (R)-(+)-glycidol, Young et al.[59] reported the synthesis of some 2',3'-dideoxy- and 2',3'-didehydro-2',3'-dideoxy-4'-thiopyrimidine nucleosides. After DIBAL-H reduction and acetylation using acetic anhydride in presence of DMAP, the thiolactone **264** was converted to the 1-O-acetyl derivative **265**. Condensation of various silylated pyrimidine bases using tin(IV) chloride or potassium nonaflate ($C_4F_9SO_3K$) provided, after removal of the silyl protecting group by fluoride cleavage, the expected α- and β-L-4'-thionucleosides of cytosine (**266** and **267**), 5-fluorocytosine (**268** and **269**), thymine (**270** and **271**), and uracil (**272** and **273**) (Scheme 36).

Phenylselenation at the 2-position of **264**, using first LiHMDS at −78 °C and then phenylselenenyl bromide,[46] led to the 2-substituted thiolactone **274** (Scheme 37). Further reduction and acetylation, as used for synthesis of **265**, provided the 1-O-acetyl-2-phenylselenenylthiolactone **275**. Condensation of the nucleic acid bases was achieved under the same conditions as described in Scheme 36 [tin(IV) chloride or potassium nonaflate], followed by oxidation by m-CPBA, which led to the elimination of PhSe, yielding the desired 2',3'-didehydro-2',3'-dideoxy nucleoside. Final removal of the silyl group using Et_4NF provided the β-L-d4-thionucleosides as cytosine (**276**), 5-fluorocytosine (**277**), thymine (**278**), and uracil (**279**) derivatives (Scheme 37).

L-2'-Deoxy-4'-thio-pyrimidine nucleosides have also been synthesized.[60,61] The condensation of various pyrimidine bases was achieved by Uenishi et al.[60] using tin(IV) chloride on the key intermediate 1-O-ethyl-5-O-acetyl-3-O-tert-butyldimethylsilyl-2-deoxy-4-thioxylofuranoside **280**. Final deprotection of silyl and acetyl groups was achieved using fluoride cleavage and methanolic ammonia, respectively, thus providing the corresponding 2'-deoxy-4'-thionucleosides **281–283** (Scheme 38).

Synthesis of 5-ethyl derivative of 2'-deoxy-4'-thiouridine has been achieved by Selwood et al.[62] The iodo-lactonization of 3-O-tert-butyldimethylsilyl-N,N-dimethyl-4,5-pentenamide **284** provided, af-

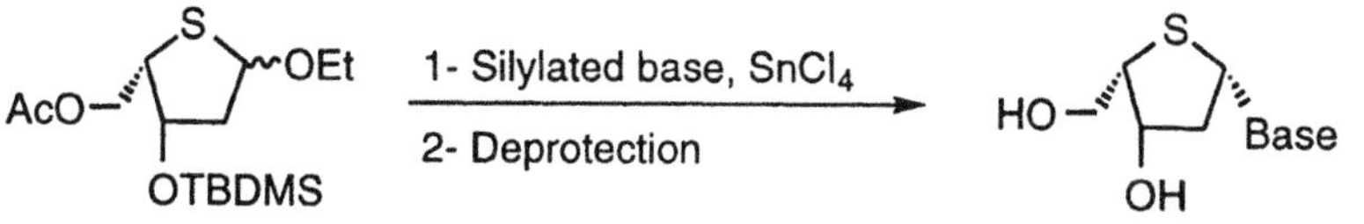

281 Base = Cytosine
282 Base = Thymine
283 Base = Uracil

Scheme 38.

ter five steps, the racemic mixture (D- and L-enantiomers) of 1,5-di-*O*-acetyl-3-*O*-*tert*-butyldimethylsily-4-thioribofuranoside **285**. Condensation with silylated 5-ethyluracil in presence of TMSOTf as catalyst, followed by deprotection using TBAF/THF and then sodium methoxide, provided the racemic mixture of 2′-deoxy-5-ethyl-4′-thiouridine (D- and L-enantiomers of α- and β-isomers). HPLC separation on chiral column resolved to the α- and β-L-thiouridine **286** and **287** (Scheme 39).

B. Two Heteroatoms Containing Nucleosides

Since the L-enantiomer of BCH-189 (3TC, Epivir®, Lamivudine) has been shown to be more potent and less toxic than its D counterpart,[63] 1,3-oxathiolanyl-L-nucleosides are considered as lead compounds. Thus, the mechanism of condensation step between nucleic acid bases and oxathiolanyl pseudosugar moiety was precisely studied.[64] Starting from a key intermediate **288**, obtained from L-gulose, Jeong et al.[65] synthesized a series of enantiomerically pure L-oxathiolanyl pyrimidine and purine nucleosides. Several cytosine (**289–300**) and uracil (**301–312**) derivatives have been obtained by coupling the corresponding silylated nucleic acid bases in presence of TMSOTf on **288**, followed by removal of silyl group using *n*-Bu_4NF in THF (Scheme 40).

Using the same coupling and deprotection methodologies, α- and β-L-purine nucleosides have been synthesized (**313–320**, Scheme 40).

Previously to the enantioselective synthesis described by Jeong et al, Hoong et al.[66] had obtained enantiomerically pure β-D- or β-L-oxathiolanyl analogs using an enzyme-mediated preparation. After *O*-acylation of its 5′-hydroxymethyl group, a racemic mixture of 3TC

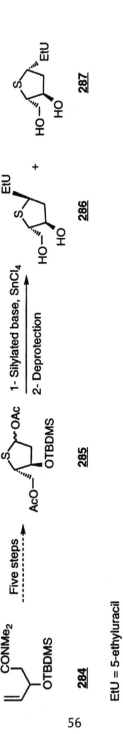

Scheme 39.

EtU = 5-ethyluracil

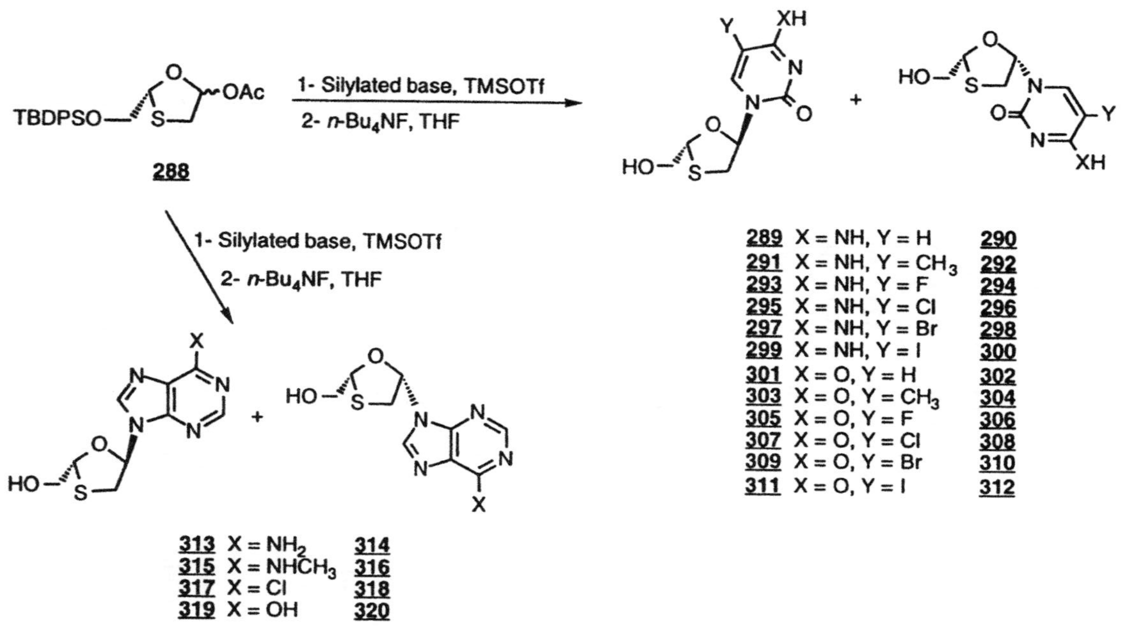

289   X = NH, Y = H   290
291   X = NH, Y = CH$_3$ 292
293   X = NH, Y = F   294
295   X = NH, Y = Cl   296
297   X = NH, Y = Br   298
299   X = NH, Y = I   300
301   X = O, Y = H   302
303   X = O, Y = CH$_3$ 304
305   X = O, Y = F   306
307   X = O, Y = Cl   308
309   X = O, Y = Br   310
311   X = O, Y = I   312

1- Silylated base, TMSOTf
2- n-Bu$_4$NF, THF

288

1- Silylated base, TMSOTf
2- n-Bu$_4$NF, THF

313   X = NH$_2$ 314
315   X = NHCH$_3$ 316
317   X = Cl   318
319   X = OH   320

Scheme 40.

57

or FTC was stereoselectively deacetylated by enzymes. These enzymes hydrolyze specifically the acyl-(+)- or (–)-β-enantiomer, leading to the exclusively corresponding β-D- or β-L-pure enantiomer. For example, the lipase CE has a specific affinity for the (+)-enantiomer, while the lipase PS 30 is specific of the (–)-isomer.

Kim et al.[67] reported the synthesis of enantiomerically pure L-dioxolanyl pyrimidine and purine nucleosides using an analogous procedure employed for the synthesis of oxathiolanyl nucleosides. Starting from the key intermediate **321** obtained from 2,3:5,6-di-O-isopropylidene-L-gulofuranose in seven steps, the synthetic route featured the coupling of various silylated bases such as N^4-benzoyl-5-substituted-cytosine (H, CH_3, F, Cl, Br, I) or thymine in presence of TMSOTf and deprotection of benzoyl group using methanolic ammonia (Scheme 41).

A series of purine dioxolanyl nucleosides has also been synthesized from the same intermediate **321**. Using silylated 6-chloropurine and TMSOTf as catalyst, the corresponding α- and β-5-O-benzoylnucleosides were prepared (**336** and **337**, respectively). Compounds **336** and **337** were treated separately with methanolic ammonia at room temperature, providing various purine derivatives such as 6-chloropurine (**338** and **339**), 6-methoxypurine (**340** and **341**) and adenine (**342** and **343**). Compounds **338–340** (α-isomers) or **339–341** (β-isomers) were obtained as an inseparable mixture. Reaction of **337** (β-isomer) with sodium methoxide in MeOH yielded nucleosides **339** and **341** as an inseparable mixture. Further treatment of this mixture by mercaptoethanol in presence of sodium methoxide provided **344** (hypoxanthine derivative), which was separated from **341** (unchanged 6-methoxy derivative), while treatment by methylamine in methanol of **349/341** yielded the N^6-methyladenine derivative **345** as the only product (Scheme 42).

Starting from silylated 2-fluoro-6-chloropurine coupled to **321** using TMSOTf as catalyst, Kim et al.[67] also synthesized a series of α- and β-2,6-disubstituted purine analogs such as 2-amino-6-chloropurine (**350** and **351**, respectively), 2-fluoro-adenine (**352** and **353**, respectively), or guanine (**354** and **355**, respectively, Scheme 43). After coupling and separation of α/β isomers, treatment of α-adduct with NH_3 in DME yielded a mixture of **346** (2-amino-6-chloropurine derivative) and **348** (2-fluoroadenine analog). The β-isomer led to the

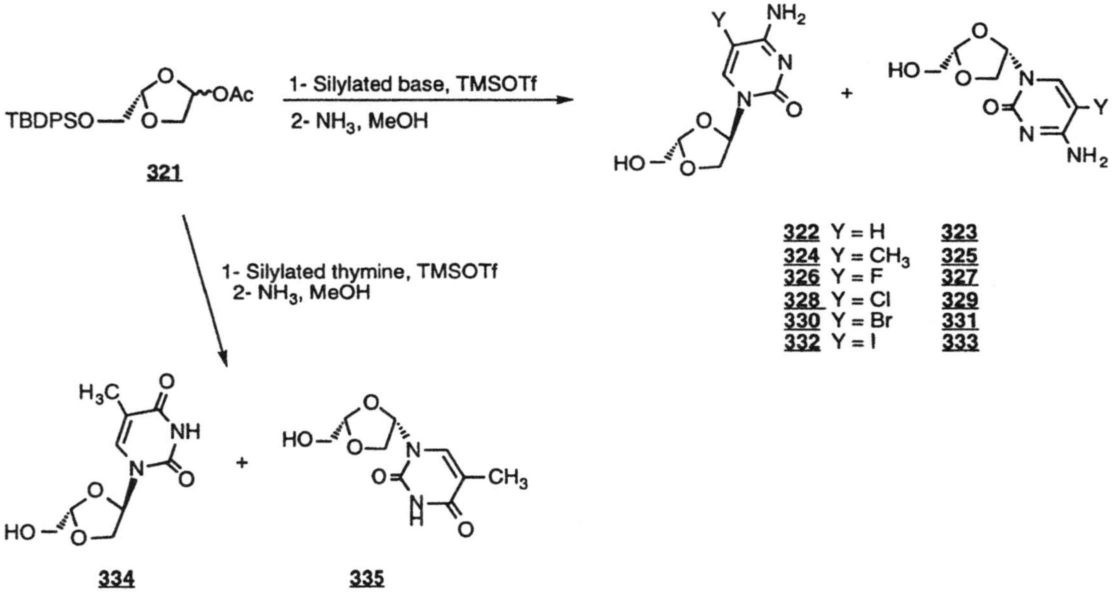

322 Y = H   **323**
324 Y = CH₃   **325**
326 Y = F   **327**
328 Y = Cl   **329**
330 Y = Br   **331**
332 Y = I   **333**

Scheme 41.

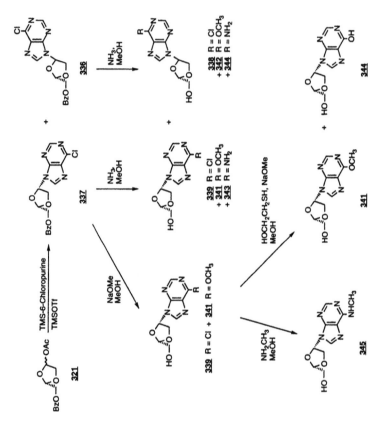

Scheme 42.

60

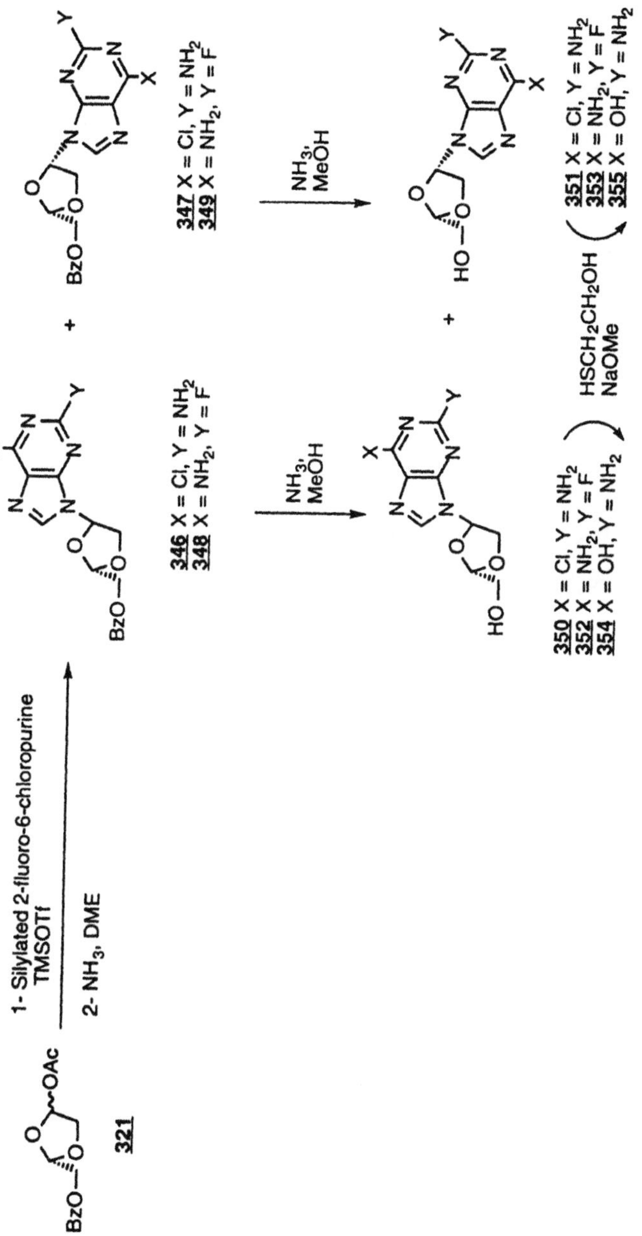
BzO
OAc
321
1- Silylated 2-fluoro-6-chloropurine
TMSOTf
2- NH3, DME
BzO
346 X = Cl, Y = NH2
348 X = NH2, Y = F
+
BzO
347 X = Cl, Y = NH2
349 X = NH2, Y = F
NH3,
MeOH
NH3,
MeOH
HO
HO
+
350 X = Cl, Y = NH2
352 X = NH2, Y = F
354 X = OH, Y = NH2
HSCH2CH2OH
NaOMe
351 X = Cl, Y = NH2
353 X = NH2, Y = F
355 X = OH, Y = NH2

Scheme 43.

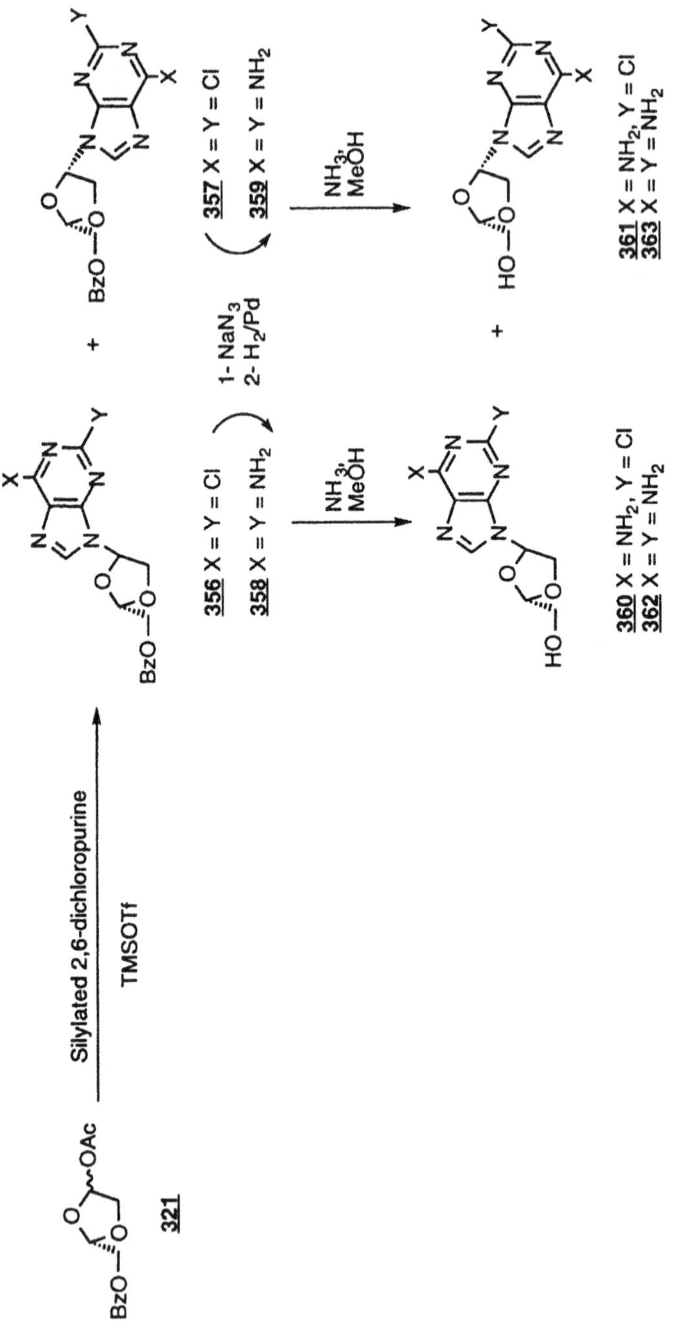

Silylated 2,6-dichloropurine

TMSOTf

321

356 X = Y = Cl
358 X = Y = NH₂

1- NaN₃
2- H₂/Pd

357 X = Y = Cl
359 X = Y = NH₂

NH₃,
MeOH

NH₃,
MeOH

360 X = NH₂, Y = Cl
362 X = Y = NH₂

361 X = NH₂, Y = Cl
363 X = Y = NH₂

Scheme 44.

mixture of corresponding β-isomers **347** and **349** when treated under the same conditions. Further treatment of these compounds with methanolic ammonia provided the deprotected compounds **350–353**. Treatment of **350** or **351** by mercaptoethanol in presence of sodium methoxide led to the guanine analog **354** or **355**, respectively (Scheme 43).

Other 2,6-disubstituted purine analogs were synthesized by Kim et al.,[67] such as 2-chloroadenine or 2,6-diaminopurine (Scheme 44). Condensation of 2,6-dichloropurine on **321** using TMSOTf provided α- and β-isomers of 2,6-dichloropurine analogs **356** and **357**, respectively. Treatment of **356** or **357** with methanolic ammonia led to the deprotected 2-chloroadenine dioxolanyl derivatives **360** or **361**, respectively. Reaction of **356** or **357** with sodium azide followed by H_2/Pd reduction provided the 2,6-diamino derivatives **358** and **359** (Scheme 44). Final deprotection with NH_3/MeOH yielded the 2,6-diaminodioxolanylpurine analogs **362** and **363**.

VII. OTHER L-HETERONUCLEOSIDES

The L-nucleosides described in this article are mirror images of the natural (D-) nucleosides, possess 2′ and/or 3′-substituent modifications (azido, fluoro, 2′,3′-didehydro, for example) or incorporate different heteroatoms in the ribose moiety. However, other "D-" and "L-" nucleosides have also been studied wherein the heteroatom of the

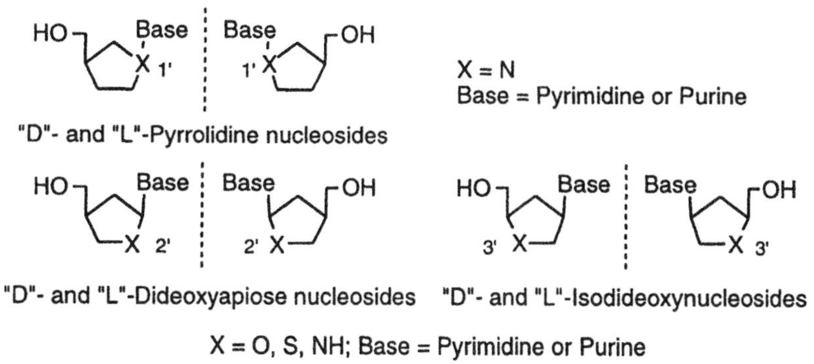

"D"- and "L"-Pyrrolidine nucleosides

X = N
Base = Pyrimidine or Purine

"D"- and "L"-Dideoxyapiose nucleosides   "D"- and "L"-Isodideoxynucleosides

X = O, S, NH; Base = Pyrimidine or Purine

Figure 8.

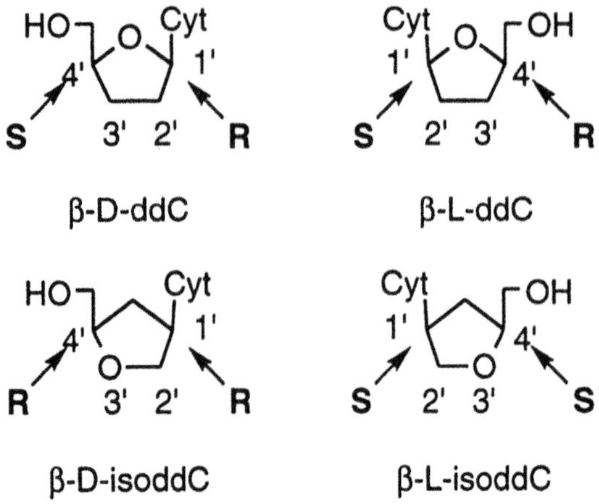

β-D-ddC   β-L-ddC

β-D-isoddC   β-L-isoddC

Figure 9.

sugar moiety is moved from its original 4′-position to the 1′-position (pyrrolidine nucleosides), 2′-position (dideoxyapiose nucleosides), or 3′-position (iso-dideoxynucleosides) (Figure 8).

Nair et al.[68] reported the antiviral activities of such D- and L-nucleosides. These nucleosides incorporated in the sugar moiety different heteroatoms such as O, S, or N and various nucleic acid bases such as pyrimidine or purine bases (Figure 8).

When the C4′ of the D-nucleosides and the L-nucleosides are in S and R configurations respectively (CH–CH$_2$OH), this same carbon in the isodideoxynucleoside structure is in an R configuration for the D-isomer and in an S configuration for the L-isomer (Comparison of D-/L-ddC and D-/L-isoddC, Figure 9).

VIII.   CONCLUSION

The synthesis of so-called L-nucleosides has provided a renewed interest in nucleoside analogs as a class of antiviral agents. Studies on the biochemistry and cellular pharmacology of these compounds have been in general quite rewarding. In particular, these compounds have different selectivity (potency and toxicity) and resistance profile when compared to their D-counterparts. In effect, a combination of the D-

and L-enantiomers of FTC (Racivir™) is currently being considered as the first rationally designed co-active modality.[9]

ACKNOWLEDGMENTS

This work was supported in part by NIH grants 1RO1-AI-41980, 1RO1-AI32351, 2R44-CA-65434, 1RO1-CA-53892, the Department of Energy grant DE-FG02-96ER62156, the Department of Veterans Affairs, and the Georgia Research Center for AIDS and HIV Infections. This review was completed on July 17, 1998 and is dedicated to Dr. William H. Prusoff.

REFERENCES

1. Prusoff, W. H.; Chen, M. S.; Fisher, P. H.; Lin, T.-S.; Shiau, G.T.; Schinazi, R.F.; Walker, J. *Pharmacol. Ther.* **1979**, *7*, 1–34.
2. Belleau, B.; Dixit, D.; Nguyen-Ba, N.; and Kraus, L. *V Internl. Conference on AIDS, Montreal (Quebec) Canada*, June 4–9, 1989.
3. Schinazi, R. F.; Chu, C. K.; Peck, A.; McMillan, A.; Mathis, R.; Cannon, D.; Jeong, L.-S.; Beach, J. W.; Choi, W.-B.; Yeola, S.; Liotta, D. C. *Antimicrob. Agents Chemother.* **1992**, *36*, 672–676.
4. Coates, J.; Cammack, N.; Kenkinson, H.; Mutton, I.; Pearson, B.; Storer, R.; Cameron, J.; Penn, C. *Antimicrob. Agents Chemother.* **1992**, *36*, 202–205.
5. Furman, P.; Painter, G. R. *Internl. Antiviral News* **1995**, *3*, 74–77.
6. Schinazi, R. F.; McMillan, A.; Cannon, D.; Mathis, R.; Lloyd, R. M.; Peck, A.; Sommadossi, J.-P.; St. Clair, M.; Wilson, J.; Furman, P. A.; Painter, G.; Choi, W.-B.; Liotta, D. C. *Antimicrob. Agents Chemother.* **1992**, *36*, 2423–2431.
7. Furman, P. A.; Davis, M.; Liotta, D. C.; Paff, M.; Frick, L. W.; Nelson, D. J.; Dornsife, R. E.; Wurster, J. A.; Wilson, L. J.; Fyfe, J. A.; Tuttle, J. V.; Miller, W. H.; Condreay, L.; Averett, D. R.; Schinazi, R. F.; Painter, G. R. *Antimicrob. Agents Chemother.* **1992**, *36*, 2686–2692.
8. Grove, K. L.; Guo, X.; Liu, S. H.; Gao, Z.; Chu, C. K.; Cheng, Y. C. *Cancer Res.* **1995**, *55*, 3008–3011.
9. Schinazi, R. F.; McMillan, A.; Lloyd, Jr., R. L.; Schlueter-Wirtz, S.; Liotta, D. C.; Chu, C. K. *Antiviral Res.* **1997**, *34*, A42.
10. Ries, S.; Wert, V.; O'Leary, N. F. D.; Nair, M. *Plant Growth Regul.* **1992**, *9*, 263–273.
11. Holý, A.; Sorm, F. *Collection Czechoslov. Chem. Comm.* **1969**, *34*, 3383.
12. Burnstock, G.; Cusack, N. J.; Hills, J. M.; Mc Kenzie, I.; Meghji, P. *Br. J. Pharmacol.* **1983**, *79*, 907.
13. Gouch, G.; Maguire, M. H. *J. Med. Chem.* **1967**, *10*, 475.
14. Cusack, N. J.; Planker, M. *Br. J. Pharmacol.* **1979**, *67*, 153.
15. Shaeffer, H. J.; Thomas, H. J. *J. Am. Chem. Soc.* **1958**, *80*, 3738.
16. Asai, M.; Hieda, H.; Shimizu, B. *Chem. Pharm. Bull.* **1967**, *15*, 1863.

17. Davoll, J.; Lowy, B. A. *J. Am. Chem. Soc.* **1951**, *73*, 1650–1655.
18. Génu-Dellac, C.; Gosselin, G.; Aubertin, A. M.; Obert, G.; Kirn, A.; Imbach, J.-L. *Antiviral Chem. Chemother.* **1991**, *2*, 83–92.
19. Génu-Dellac, C.; Gosselin, G.; Puech, F.; Henry, J. C.; Aubertin, A. M.; Obert, G.; Kirn, A.; Imbach, J.-L. *Nucleosides Nucleotides* **1991**, *10*, 1345–1376.
20. Holy, A. *Collection Czechoslov. Chem. Comm.* **1972**, *37*, 4072.
21. Jansons, J.; Maurinsh, Y.; Lidaks, M. *Nucleosides Nucleotides* **1995**, *14*, 1709.
22. Gosselin, G.; Bergogne, M.-C.; Imbach, J.-L. *J. Heterocyclic. Chem.* **1993**, *30*, 1229–1233.
23. Lin, T.-S.; Luo, M. Z.; Liu, M. C. *Tetrahedron* **1995**, *51*, 1055–1068.
24. Holý, A. *Tetrahedron Lett.* **1971**, *2*, 189–192.
25. Robins, M. J.; Khwaja, T. A.; Robins, R. K. Purine nucleosides. *J. Org. Chem.* **1970**, *35*, 636–639.
26. Hoffer, M. *Chem. Ber.* **1960**, *93*, 2777–2781.
27. Humoller, F. L. *Methods Carbohyd. Chem.* **1963**, *1*, 83–88.
28. Robins, M. J.; Robins, R. K. *J. Amer. Chem. Soc.* **1965**, *81*, 4934–4940.
29. Robins, M. J.; Robins, R. K. *J. Org. Chem.* **1969**, *34*, 2160–2163.
30. Spadari, S.; Maga, G.; Focher, F.; Ciarrocchi, G.; Mansevigi, R.; Arcamone, F.; Capobianco, M.; Carcuro, A.; Colonna, F.; Iotti, S.; Garbesi, A. *J. Med. Chem.* **1992**, *35*, 4214–4220.
31. Coderre, J. A.; Santi, D. V.; Matsuda, A.; Watanabe, K. A.; Fox, J. J. *J. Med. Chem.* **1983**, *26*, 1149–1152.
32. Codington, J. F.; Doerr, I. L.; Fox, J. J. *J. Org. Chem.* **1964**, *29*, 558–564.
33. Harada, K.; Matulic-Adamic, J.; Price, R. W.; Schinazi, R. F.; Watanabe, K. A.; Fox, J. J. *J. Med. Chem.* **1987**, *30*, 226–229.
34. Xiang, Y.; Kotra, L.; Chu, C. K.; Schinazi, R. F. *Bioorg. Med. Chem. Lett.* **1995**, *5*, 743–748.
35. Ma, T.; Pai, S. B.; Zhu, Y. L.; Lin, J. S.; Shanmuganathan, K.; Du, J.; Wang, C.; Kim, H.; Newton, M. G., Cheng, Y.-C,; Chu, C. K. *J. Med. Chem.* **1996**, *39*, 2835–2843.
36. Chu, C. K.; Ma, T.; Shanmuganathan, K.; Wang, C.; Xiang, Y. J.; Pai, S. B.; Yao, G. Q.; Sommadossi, J.-P.; Cheng, Y. C. *Antimicrob. Agents Chemother.* **1995**, *39*, 979–981.
37. Mathé, C.; Gosselin, G.; Bergogne, M. C.; Aubertin, A. M.; Obert, G.; Kirn, A.; Imbach, J.-L. *Nucleosides Nucleotides* **1995**, *14*, 549–550.
38. Lin, T.-S.; Luo, M. Z.; Liu, M. C.; Pai, S. B.; Duschman, G. E.; Cheng, Y. C. *J. Med. Chem.* **1994**, *37*, 798–803.
39. Lin, T. S.; Luo, M. Z.; Liu, M. C. *Tetrahedron* **1995**, *51*, 1055–1068.
40. Lin, T.-S.; Luo, M. Z.; Liu, M. C. *Tetrahedron Lett.* **1994,** *35*, 3477–3480.
41. (a) Gosselin, G.; Mathé, C.; Bergogne, M. C.; Aubertin, A. M.; Kim, A.; Sommadossi, J.-P.; Schinazi, R. F.; Imbach, J.-L. *Nucleosides Nucleotides* **1995**, *14*, 611–617. (b) Gosselin, G.; Mathé, C.; Bergogne, M.-C.; Aubertin, A.-M.; Kirn, A.; Schinazi, R. F.; Sommadossi, J.-P.; Imbach, J.-L. *Sciences de la Vie* **1994**, *317*, 85–89. (c) Schinazi, R. F.; Gosselin, G.; Faraj, A.; Korba, B. E.;

Liotta, D. C.; Chu, C. K.; Mathé, C.; Imbach, J.-L.; Sommadossi, J.-P. *Antimicrob. Agents Chemother.* **1994,** *38,* 2172–2174.

42. Van Draanen, N. A.; Koszalka, G. *Nucleosides Nucleotides* **1994,** *13,* 1679–1693.

43. Mansuri, M. M.; Farina, V.; Starrett Jr, J. E.; Benigni, D. A.; Brankovan, V.; Martin, J. C. *Bioorg. Med. Chem. Lett.* **1991,** *1,* 65–68.

44. Lin, T.-S.; Luo, M. Z.; Zhu, J.-L.; Liu, M. C.; Zhu, Y. L.; Dutschman, G. E.; Cheng, Y. C. *Nucleosides Nucleotides* **1995,** *14,* 1759–1783.

45. Bolon, P.; Wang, P.; Chu, C. K.; Gosselin, G.; Boudou, V.; Pierra, C.; Mathé, C.; Imbach, J.-L.; Faraj, A.; el Alaoui, M. A.; Sommadossi, J.-P.; Pai, S. B.; Zhu, Y. L.; Lin, J. S.; Cheng, Y.-C.; Schinazi, R. F. *Bioorg. Med. Chem. Lett.* **1996,** *6,* 1657–1662.

46. Beach, J. W.; Kim, H. O.; Jeong, L. S.; Nampalli, S.; Islam, Q.; Ahn, S. K.; Babu, J. R.; Chu, C. K. *J. Org. Chem.* **1992,** *57,* 3887–3894.

47. El Alaoui, A. M.; Faraj, A.; Pierra, C.; Boudou, V.; Johnson, R.; Mathé, C.; Gosselin, G.; Korba, B. E.; Imbach, J.-L.; Schinazi, R. F.; Sommadossi, J.-P. *Antiviral Chem. Chemother.* **1996,** *7,* 276–280.

48. Burns, C. L.; St Clair, M. H.; Frick, L. W.; Spector, T.; Averett, D. R.; English, M. L.; Holmes, T. J.; Krenitsky, T. A.; Koszalka, G. W. *J. Med. Chem.* **1993,** *36,* 378–384.

49. Lee, C. S.; Du, J.; Chu, C. K. *Nucleosides Nucleotides* **1997,** *15,* 1223–1236.

50. Xiang, Y.; Cavalcanti, S.; Chu, C. K.; Schinazi, R. F.; Pai, S. B.; Zhu, Y. L.; Cheng, Y. C. *Bioorg. Med. Chem. Lett.* **1995,** *5,* 877–880.

51. Wengel, J.; Lau, J.; Pedersen, E. B.; Nielsen, C. M. *J. Org. Chem.* **1991,** *56,* 3591–3594.

52. Czernecki, S.; Le Diguarher, T. *Synthesis* **1991,** 683–686.

53. Sugimura, H.; Osumi, K.; Yamazaki, T.; Yamaya, T. *Tetrahedron Lett.* **1991,** *32,* 1813–1816.

54. Gould, J. H.; Mann, J. *Nucleosides Nucleotides* **1997,** *16,* 193–213.

55. Lin, T. S.; Luo, M. Z.; Liu, M. C.; Zhu, Y. L.; Gullen, E.; Dutschman, G. E.; Cheng, Y.-C. *J. Med. Chem.* **1996,** *39,* 1757–1759.

56. Graciet, J.-C.; Shi, J.; Schinazi, R. F. *Nucleosides Nucleotides* **1998,** *17,* 711–722

57. Schinazi, R. F.; Goudgaon, N. M.; Fulcrand, G.; El Kattan, Y.; Lesnikowski, Z.; Ullas, G. V.; Moravek, J.; Liotta, D. C. *Int. J. Radiation Oncol. Biol. Phys.* **1994,** *28,* 1113–1120.

58. El Kattan, Y.; Goudgaon, N. M.; Fulcrand, G.; Liotta, D. C.; Schinazi, R. F. *Current Topics in the Chemistry of Boron*; Kabalka, G. W. Ed.; The Royal Society of Chemistry: England, 1994, pp. 181–184.

59. Young, R. J.; Shaw-Ponter, S.; Thomson, J. B.; Miller, J. A.; Cumming, J. G.; Pugh, A. W.; Rider, P. *Bioorg. & Med. Chem. Lett.* **1995,** *5,* 2599–2604.

60. Uenishi, J.; Takahashi, K.; Motoyama, M.; Akashi, H.; Sasaki, T. *Nucleosides Nucleotides* **1994,** *13,* 1347–1361.

61. Tiwari, K. N.; Montgomery, J. A.; Secrist III, J.A. *Nucleosides Nucleotides* **1993**, *12*, 841–846.
62. Selwood, D. L.; Carter, K.; Young, R. J.; Jandu, K. S. *Bioorg. & Med. Chem. Lett.* **1996**, *6*, 991–994.
63. Doong, S. L.; Tsai, C. H.; Schinazi, R. F.; Liotta, D. C.; Cheng, Y.-C. *Proc. Natl. Acad. Sci. USA* **1991**, *88*, 8495–8499.
64. Choi, W.-B.; Wilson, L. J.; Yeola, S.; Liotta, D. C.; Schinazi, R. F. *J. Amer. Chem. Soc.* **1991**, *113*, 9377–9379.
65. Jeong, L. S.; Schinazi, R. F.; Beach, J. W.; Kim, H. O.; Nampalli, S.; Shan-muganathan, K.; Alves, A. J.; McMillan, A.; Chu, C. K.; Mathis, R. *J. Med. Chem.* **1993**, *36*, 181–195.
66. Hoong, L. K.; Strange, L. E.; Liotta, D. C.; Koszalka, G. W.; Burns, C. L.; Schinazi, R. F. *J. Org. Chem.* **1992**, *57*, 5563–5565.
67. Kim, H. O; Schinazi, R. F.; Shanmuganathan, K.; Nampalli, S.; Jeong, L. S.; Beach, J. W.; Nampalli, S.; Cannon, D. L.; Chu, C. K. *J. Med. Chem.* **1993**, *36*, 519–528.
68. Nair, V.; Jahnke, T. S. *Antimicrob. Agents Chemother.* **1995**, *39*, 1017–1029.

ORALLY BIOAVAILABLE ACYCLIC NUCLEOSIDE PHOSPHONATE PRODRUGS:
ADEFOVIR DIPIVOXIL AND BIS(POC)PMPA

Murty N. Arimilli, Joseph P. Dougherty,
Kenneth C. Cundy, and Norbert Bischofberger

Advances in Antiviral Drug Design
Volume 3, pages 69–91.
Copyright © 1999 by JAI Press Inc.
All rights of reproduction in any form reserved.
ISBN: 0-7623-0201-1

I. INTRODUCTION

Acyclic nucleoside and nucleotide analogs have shown substantial clinical utility against a variety of viral infections. Considerable excitement has been generated by biological data on nucleoside and nucleotide analogs with potent antiviral and antineoplastic effects.[1] In the nucleoside series (Figure 1), acyclovir (ACV, **1**) is approved in the United States and Europe for herpes simplex[2] and herpes zoster infections[3] and famciclovir (**4**), a prodrug of penciclovir (**3**), is approved for acute herpes zoster infections.[4] Ganciclovir (**2**) is in use for cytomegalovirus (CMV) infections and prophylaxis in immunocompromised patient populations.[5,6] Other nucleosides (Figure 1) ap-

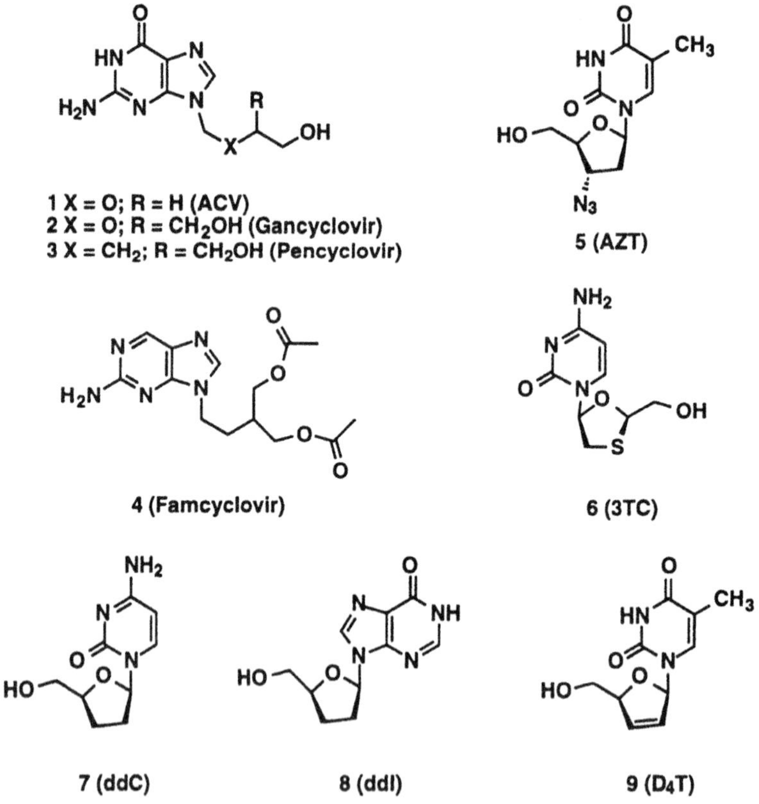

1 X = O; R = H (ACV)
2 X = O; R = CH$_2$OH (Gancyclovir)
3 X = CH$_2$; R = CH$_2$OH (Penicyclovir)

5 (AZT)

4 (Famcyclovir)

6 (3TC)

7 (ddC)

8 (ddI)

9 (D$_4$T)

Figure 1.

proved are AZT (**5**)[7], 3TC (**6**)[8], ddC (**7**)[9], ddI (**8**),[10] and D_4T (**9**).[11] In addition, a number of nucleoside analogs are in earlier stages of development.

Acyclic nucleotide analogs (Figure 2) also show potent antiviral activity. Cidofovir (HPMPC, **10**), an unusually broad-spectrum antiviral agent[12] with activity against herpes viruses, adenovirus, and papillomavirus, has been approved for therapy of CMV retinitis. Intravenous adefovir (PMEA, **11**), has completed phase I/II trials for the treatment of HIV infection.[13] PMPA (**12**) has shown potent and selective activity against human immunodefficiency virus (HIV) and other retroviruses.[14] Recently, it was shown that PMPA was able to completely prevent SIV infection in macaques even when treatment was initiated as late as 24 h after inoculation.[15] In a phase I/II human clinical study, PMPA given intravenously, was safe and well tolerated and caused a 1.1 log reduction of plasma HIV-RNA levels after only eight doses.[16]

Nucleosides and nucleotides usually require conversion to their triphosphate derivative for their biological activity. These active triphosphates target viral or cellular polymerases, inhibiting or terminating the growing polynucleotide chain. Thus, nucleosides have to undergo three distinct phosphorylation steps, consecutively to the mono-, di-, and triphosphates. In this sequence, the first phophorylation requires specific enzymes and can be rate limiting, whereas phosphorylation to the di- and triphosphates is carried out by a number of unspecific cellular kinases. This anabolic activation is advantageous under some circumstances, as for example when ACV is phos-

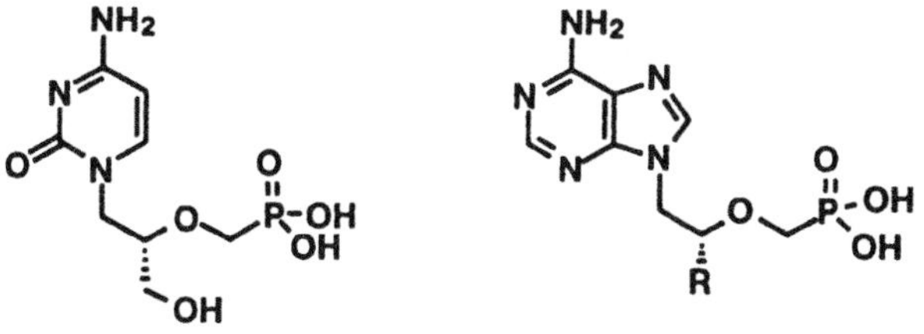

10 (Cidofovir, HPMPC)

11 R = H (Adefovir, PMEA)
12 R = CH₃ (PMPA)

Figure 2.

phorylated by a virally encoded thymidine kinase (TK),[17,18] and then further elaborated to the triphosphate by cellular kinases.[19] This viral activation is responsible for much of the selectivity and low toxicity of ACV, as uninfected cells do not activate the drug.[20]

The corresponding disadvantage of the specific viral activation of nucleoside derivatives is that it can easily lead to a pathway for the selection of resistant viral mutants. Some potentially active nucleosides are never substrates for an appropriate enzyme, and therefore may fail to show any biological activity. Mutant viruses or cancer cells that can manage not to phosphorylate the antiviral drug will have a selective advantage over wild type under conditions of drug therapy. The most common ACV-resistant mutants of HSV, for example, are due to inactivation or alteration of the viral TK.

Ganciclovir is also an excellent substrate for the HSV TK,[21] and is active against the virus.[22] CMV does not encode a TK, but the UL97 gene product can phosphorylate GCV.[23] This phosphorylation is required for GCV's activity against CMV, and the inability of ACV to serve as a substrate of UL97 is one reason why ACV is much less active against CMV.[24] Since a virally encoded enzyme is required for GCV activity, it also opens a mechanism for resistance. Indeed, ganciclovir-resistant clinical isolates of CMV often have mutant UL97 gene products that do not phosphorylate GCV.[25] Some GCV-resistant clinical isolates also map to the CMV polymerase gene.[26]

In the case of nucleosides that rely on activation by endogenous cellular enzymes, there is the risk that target cells do not express the needed enzyme at the required time in the required tissue. AZT needs phosphorylation by an endogenous TK, which is present only at very low levels in resting cells.[27] HIV target cells such as macrophages which reside in the G_0 phase of the cell cycle phosphorylate AZT inefficiently.

Nucleotide drugs, in contrast, are chemically provided with their initial phosphorylation, and so sidestep any enzymatic requirement for this first activation step. The phosphonomethylether class of nucleotide analogs has shown particular promise since their discovery by Holy and De Clercq.[1] This class has a nonhydrolyzable linkage of the nucleoside moiety to the phosphonate that is isosteric with an ordinary phosphomonoester. Simple phosphomonoesters of nucleoside drugs are typically good substrates for phosphatases and are degraded to the parent nucleoside in biological systems.[28]

Phosphonomethylether nucleosides have unusual pharmacokinetic properties. They are very slow to pass biological membranes presumably due to the negative charge on the phosphonate moiety. This charge also entails the low oral bioavailability of the native phosphonates;[29] for example < 5% for adefovir and cidofovir in rats, and similarly low numbers in monkeys[30] and humans.[31] Nucleoside phosphonates are not taken up into cells by the nucleotidase/nucleoside transport mechanism used for ordinary nucleotides,[32] yet they still penetrate by an alternative active transport mechanism[33] or a mixed mechanism.[34] This mechanism, though, is not very efficient, and nucleoside phosphonates poorly penetrate cells. In an uptake experiment with radio-labeled cidofovir, only 1% was absorbed by cells.[34] However, once in cells, they tend to have long half-lives, in part due to their biochemical modification to nondiffusing choline esters and phosphate anhydrides. Adefovir has an *in vitro* intracellular half-life of 16–18 h,[35] and cidofovir's is even longer.[36] This slow clearance is reflected in long *in vivo* half-lives, as well (~36 h for cidofovir in monkeys as measured with radioactive drug). Thus, these acyclic nucleotides represent active parenteral agents whose oral bioavailability is limited. This limitation can successfully be addressed by utilizing a prodrug strategy.

II. NUCLEOTIDE PRODRUGS

A satisfactory prodrug must simultaneously fulfill a number of demanding criteria. It must be chemically stable to formulation, prolonged storage, and the acidic environment of the stomach. Biologically, it should easily permeate the intestinal wall, and revert to give the parent drug in the systemic circulation (or alternatively, inside target cells,[37] see Figure 3).

For phosphonomethylether nucleosides, a solution to these difficulties is to neutralize the charges on phosphorous by derivatization to esters or amidates. In addition to giving a neutral molecule, the particular derivative can be chosen to optimize other properties of the molecule, including solubility and lipophilicity.

A prodrug approach presents particular challenges for phosphonate drugs, since the two positions to be protected have significantly

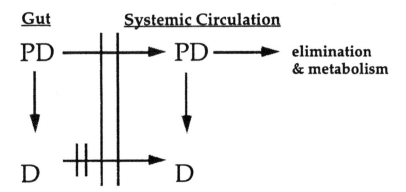

Figure 3. Prodrug (PD) used to increase oral bioavailability of parent drug (D): PD should be stable in the gut, permeate across the intestinal wall, and, once in the systemic circulation, should be efficiently converted to D.

different properties. The two different ester positions, while identical in PMEA or other achiral phosphonates, have quite different sequential reactivities.

The phosphonate diester **I** (Figure 4) has a neutral phosphorous center which is susceptible to hydrolysis chemically or enzymatically. The product **II** of this reaction is a monoanion at physiological pH, and is therefore resistant to further chemical attack and often to enzymatic attack as well (i.e. formation of **III** is usually slower). Unless a specific enzyme, such as a phosphodiesterase[28] can catalyze this reaction, it is chemically more difficult to place the nucleophile in close proximity to the anionic substrate **II**. Alternatively, a promoiety may be chosen as shown in **IV** or **V** which relies on the expulsion of a phosphonate anion, breaking the O–R bond rather than the O–P bond. The monoanion **V** is an excellent leaving group, with a pK_a below 2, but the dianion **VI** is more basic, with a pK_a near neutrality. This renders the formation of **V** is much easier than **VI**. In addition, the anion generated by either path may often lower the affinity of the new monoester (**II** or **V**) for any activating enzyme. In the absence of specific enzymes, formation of **III** is more sluggish than **VI** from a chemical perspective. Prodrugs that take advantage of

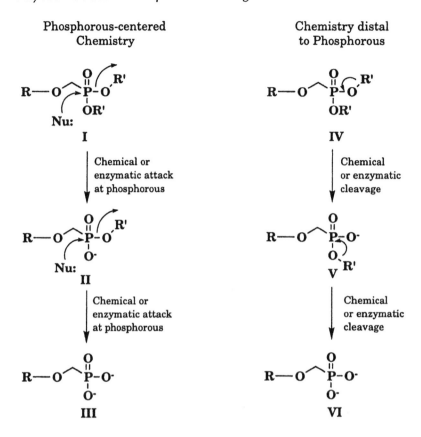

Figure 4. Mechanisms for conversion of phosphonate ester prodrugs.

chemistry not at phosphorous may therefore have a better chance *a priori*, although the exact mechanisms that operate *in vivo* have usually not been rigorously demonstrated.

Thus, the promoieties considered must ultimately be unusually versatile in their initial stability and eventual reactivity. Finding molecules that are sufficiently reactive in the final step, while maintaining good stability as the prodrug diester, limits the scope of possible functionality. Some more general reviews of these issues have recently been published.[38,39]

III. ASSAY SYSTEMS TO EVALUATE NUCLEOTIDE PRODRUGS

The most reliable predictors of ultimate human bioavailability of drugs have been comparisons of bioavailabilities in several animal species. Since animal pharmacokinetic experiments are laborious and expensive, there is considerable incentive to find benchtop experiments that will predict the results of the animal models and the actual human clinical situation. Ideally, this allows screening of a large number of derivatives made on a small scale to find those that show the most promise for *in vivo* experiments. Other than a simple screening process, the *in vitro* experiments may help deconvolute the complicated, composite process of drug absorption, distribution, metabolism, and excretion.

A variety of *in vitro* screens have been used for the evaluation of potential prodrugs. Perhaps the simplest is chemical stability in aqueous solution over a broad pH range. Prodrugs require a minimum chemical stability to meet the requirements of formulation, and shelf-life, and to be able to survive the conditions of the digestive tract. Furthermore, prodrugs have to exhibit a minimum amount of aqueous solubility for oral absorption. Also octanol-water partition coefficients (log P) are useful as a measure for the lipophilicity of a prodrug and its ability to cross biological membranes. In the case of phosphonate prodrugs, compounds bearing log P values in the range +1.0 to +2.5 exhibited appreciable permeability *in vitro* and *in vivo*.

Biochemical data were also acquired to attempt to predict the metabolism and transport of prodrugs. Tissue fluids and homogenates were selected to approximate the various conditions that would be encountered by a prodrug before and during absorption. Rat and human tissues were used in an attempt to calibrate a rat model for preliminary *in vivo* data.[42] The selected models were intestinal wash (IW), intestinal and liver homogenates (IH, LH), and plasma. In addition, a Caco-2 (human colon adenocarcinoma) cell monolayer diffusion apparatus was employed to screen prodrugs for intestinal permeability.[40,41,42] In this apparatus permeation coefficients (k_p) are measured based on the rate of diffusion of a prodrug through a Caco-2 cell monolayer. The prodrugs bearing permeation coefficients (k_p) values $>10 \times 10^{-6}$ cm/s, in general are considered good candidates for exhibiting enhanced permeability in animal models. However, due to the complex mechanism of oral absorption no single assay is completely predictive of what is observed *in vivo*.

IV. ADEFOVIR PRODRUGS

In the case of adefovir, several publications have appeared on candidate prodrugs.[44,45,46] The prodrugs evaluated have been structurally diverse , including alkyl, and aryl esters (Figure 5, 13a–e), amidates (Figure 5, 14a–d), and acyloxyalkyl esters (see Figure 6, 16a–d).

The prodrug candidates were prepared either by alkylation of the phosphonate anion with an electrophilic promoiety, or the phosphonate was activated (to e.g. a dichloridate), followed by nucleophilic displacements with amines or alcohols. Alternatively, Mukaiyama condensation[43] was used. The details of the chemistry can be found in the particular references.

In many cases, the promoieties are conceptually related to ones that have been first used for carboxylate-containing drugs.[47] This has a chemical logic, since the typical carboxylate pK_a of 4–5 is intermediate between the two pK_as of a methylene phosphonate (<2 and >6). Many biochemical mechanisms, of course, will be highly specific for carboxylic acids and unavailable for phosphonates. The converse may also be true.

The simple alkyl diesters 13a–c showed reasonable aqueous solubility (~10 mg/mL), and good chemical stability, but were not converted to adefovir in rat or human tissue samples (Table 1). These *in vitro* properties were also reflected in *in vivo* animal models,[44,48] since 13a was well absorbed by rats (42%), but was not converted to adefovir.

On the other hand, the bis-aryl esters 13d and 13e were rapidly transported in the Caco-2 model ($k_p = > 30 \times 10^{-6}$ cm/s). They were chemically less stable at high pH than the dialkyl esters, but were adequately stable at physiological pH and stable to rat intestinal wash. They were, however, converted to adefovir in liver homogenates (Table 1). Thus, they seemed good candidates for prodrugs, with a scenario of good absorption followed by hepatic conversion to adefovir. Consequently, 13d and 13e were examined in a male Sprague-Dawley rat model[48] (Table 2). 13d was moderately absorbed as the

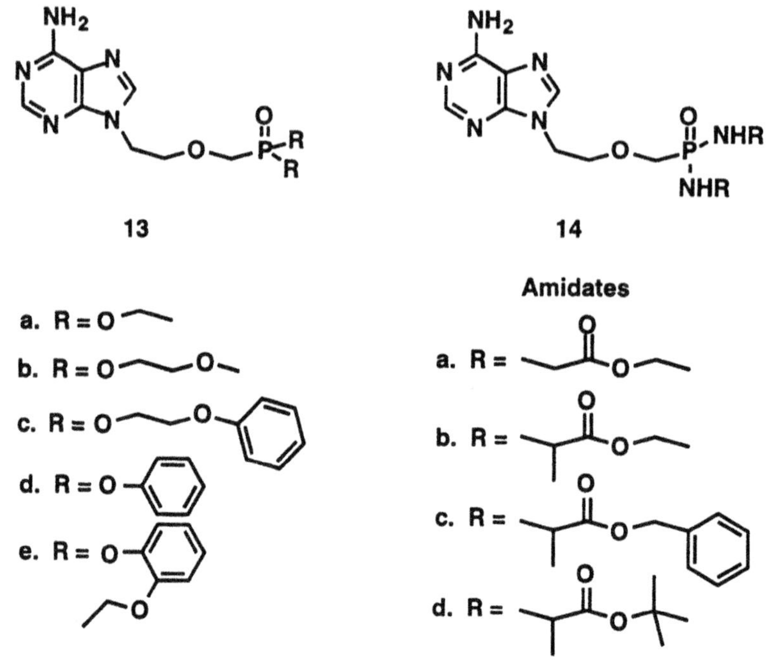

Figure 5.

Table 1. In Vitro Screens of Adefovir Prodrugs

Compound	Chem. Stability at 40 °C (h)		Log P	Biological Stability $t_{1/2}$ (min)					Caco-2
	pH = 2.0	pH = 7.4	pH = 7.4	IW Rat	IH Rat	Plasma Human	Plasma Rat	LH Human	$Kp \times 10^{-6}$ (cm/sec)
13a	> > 60	> 60	−0.26	> 60	> 60	>60	> 60	—	—
13b	> > 60	> > 60	−1.2	> 60	> 60	43	> 60	—	—
13c	> 60	> > 60	1.3	> 60	> 60	> 60	> 60	> 60	< 1.0
13d	42.6[a]	1.2[a]	1.4	> 60	> 60	> 60	> 60	60	33
13e	> 60	29	1.4	> 60	> 60	> 60	> 60	10	58
14a	5	5	—	> 60	> 60	> 60	< 10	> 60	1.8
14b	6	27	—	> 60	42	> 60	< 10	30	1.2
14c	7	34	2.1	12	1.5	> 60	40	< 1	2
14d	10	34	1.53	> 60	> 60	> 60	< 10	60	1.9
16a	14.6[a]	2.3[a]	1.96	50	< 1	3	< 10	< 1	8
16d	> 60	14	2.8	> 60	< 1	33	19	8	< 1

Note: [a]Stability at 50°C.

prodrug (22%), but poorly converted to adefovir (2.8%). In contrast, **13e** was well absorbed (51% total), and efficiently converted to adefovir (44.3%). The balance was found as several metabolites, most notably adenine acetic acid **15**[48b] (Figure 6). In this case the *in vitro* screens correctly predicted the absorption and conversion, but were unable to anticipate metabolism. Similarly, in cynomolgus monkeys both **13d** and **13e** were well absorbed and efficiently converted to adefovir (Table 2). Aryl esters of other phosphonates have also been successful as prodrugs, *vide infra.*

The alkyl phosphoramidates **14a–d**[42] were not very soluble, were poorly transported in the Caco-2 system, and tended to have limited chemical stability at acidic pH (Table 1).

Perhaps the most successful series of PMEA prodrugs to date are the acyloxyalkyl esters (Figure 6). These may be thought of as acyl-protected versions of the unstable hydroxyalkyl phosphonates. When chemically or enzymatically deacylated (i.e. chemistry distal to phosphorus, Figure 4), the hydroxyalkyl phosphonate quickly decomposes to release the phosphonic acid and the corresponding aldehyde or ketone.

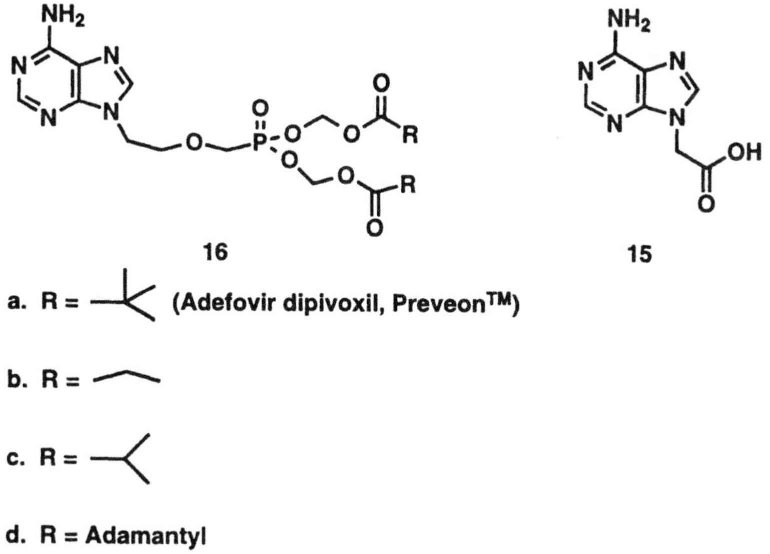

16   **15**

a. R = ⌇ (Adefovir dipivoxil, Preveon™)

b. R = ⌇

c. R = ⌇

d. R = Adamantyl

Figure 6.

A variety of acyloxyalkyl derivatives were prepared in an effort to optimize chemical and enzymatic properties. The acyloxymethyl esters **16a–d**, where R = *t*-butyl, ethyl, isopropyl, and adamantyl were examined. These compounds are lipophilic but poorly soluble. They are all unstable at basic pH and are rapidly degraded by the tissue homogenates, but they are stable to intestinal wash (Table 1). This again bodes well for their eventual absorption and conversion, although it is noteworthy that the degradation observed *in vitro* is only to the monoesters, with the subsequent conversion to adefovir being much slower. As will be seen below, this result again counsels caution in the extrapolation of bench results. When these prodrugs were tested for oral bioavailabilities in male Sprague-Dawley rats,[44,48] they gave significant improvement over the bioavailability of adefovir, with values of 17% for adefovir dipivoxil (**16a**), and 15% for the propionyl (**16b**) and isobutyryl (**16c**) compounds.[44] Subsequent experiments found the oral bioavailability of adefovir dipivoxil (**16a**) to be 42.6% in rats[48] and 27% in cynomolgus monkeys (Table 2).[50] The oral bioavailability (**16a**) in mice was found to be 53%.[49]

The analog of adefovir dipivoxil (**16a**), where R = adamantyl (**16d**), was an interesting case since this molecule was not transported across the Caco-2 cells (Table 1) but was nonetheless 19% orally bioavailable in cynomolgus monkeys.[50]

Of all the prodrugs tested, adefovir dipivoxil (**16a**) was the most satisfactory on a variety of counts: it is well-absorbed p.o. (in both rats and monkeys, Table 2) and metabolized very rapidly to release ade-

Table 2. Oral Bioavailability of Adefovir Prodrugs in Cynomolgus Monkeys and Rats

Compound	% Oral Bioavailability in Cynomolgus Monkeys		% Oral Bioavailability in Rats	
	Total Prodrug Absorbed	As Adefovir	Total Prodrug Absorbed	As Adefovir
adefovir (11)	4 ± 1	4 ± 1	6.92	6.92
13d	50	39	22	2.8
13e	49	46 ± 4	50.7	44.3
16a	27	27	42.6	42.6
16d	—	19	—	—

fovir, and apparently no other metabolites; it is a single compound with adequate if not excellent chemical stability; and it is preparable in good yield and reliable in formulation. Based on these favorable properties, adefovir dipivoxil was selected for further development.

V. PMPA PRODRUGS

In the case of adefovir dipivoxil, the pivaloyloxyalkyl ester moiety generates pivalic acid during the release of the parent drug, which at high doses can potentially decrease serum carnitine levels[51] (*vide infra*). Therefore, alkyl methyl carbonates, a new class of prodrugs of PMPA that do not liberate pivalic acid were designed for evaluation along with acyloxyalkyl esters. Alkyl ethyl carbonate prodrugs of carboxylic acids, such as bacampicilin are known in the literature.[52] However, very limited information is available on the use of alkyl methyl carbonate promoieties for phosphates or phosphonates.

The synthesis of the carbonate prodrugs **17a–g** (Figure 7) was achieved by the alkylation of PMPA with chloromethyl alkyl carbon-

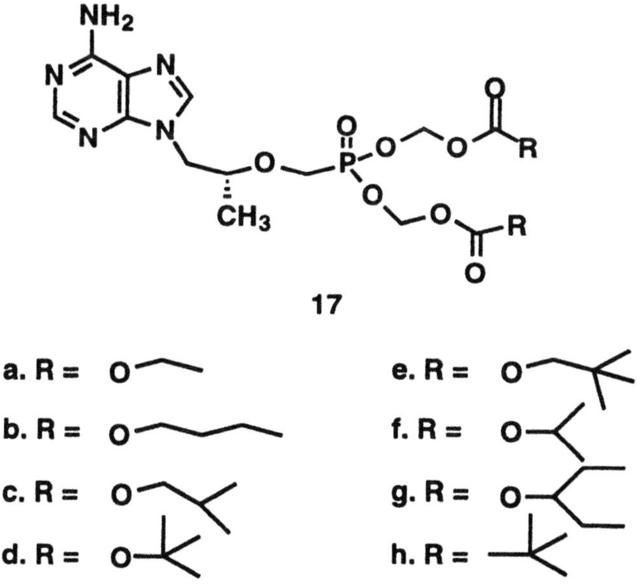

17

a. R =   O⌒

b. R =   O⌒⌒

c. R =   O⌒<

d. R =   O�

e. R =   O⌒<

f. R =   O—<

g. R =   O—<

h. R =   —<

Figure 7.

ate reagents.[53a] Bis(POM)PMPA **17h** was synthesized by the alkylation of PMPA with chloromethyl pivalate in the presence of triethylamine.[53]

The chemical stability at pHs 2.2 and 7.4, and the biological stability of these prodrugs in dog plasma, were evaluated (Table 3).[53] For comparison, the half-lives of adefovir dipivoxil **16a** at pH 2.0 and 7.4 (40 °C) were 46 h and 6 h, respectively, and adefovir dipivoxil is rapidly hydrolyzed ($t_{1/2}$ < 10 min) in human and rat plasma.

The kinetics for the degradation of all the PMPA prodrugs studied could be fit into a first order exponential decay. The phosphonate monoester was the major observed degradation product within the pH range studied. Except for **17d**, the chemical stability of the carbonate prodrugs **17a–g** (half lives of 6–9 h) are similar to adefovir dipivoxil **16a** at pH 7.4, 37 °C. At pH 2.2 they have greater stability (>150 h) except for **17d**, which showed half-life of 0.4 h at both pH values. Bis(POM)PMPA **17h** exhibits half-lives of 14 h and 68 h at pHs 7.4 and 2.2, respectively.

In the presence of dog tissue homogenates[54] (Table 3), all the carbonate prodrugs **17a–g** are readily hydrolyzed ($t_{1/2}$ < 30 min, except **17f**) to the corresponding monoester of PMPA. PMPA prodrugs **17a**, **17f** [Bis(POC)PMPA], **17g**, and **17h** [Bis(POM)PMPA] all exhibit increased stability to intestinal homogenate and to plasma rela-

Table 3. *In Vitro* Screens of PMPA Prodrugs

Compound	Chem. Stability at 37 °C (h)		Log P	Biological Stability $t_{1/2}$ (min)			Oral Bioavailability
	pH = 2.2	pH = 7.4	pH = 7.4	Dog Intestinal Homogenate	Dog Plasma	Dog Liver Homogenate	% as PMPA
17a	—	7	0.6	23.3	16.6	< 5	30
17b	> 150	6	2.6	< 5	< 5	< 5	18
17c	> 150	9	2.0	15	< 5	< 5	20.8
17d	0.4	0.4	1.9	26.6	21.2	14.9	30.7
17e	> 150	6	3.2	< 5	< 5	< 5	16
17f	> 150	8	1.3	52.6	20.5	< 5	30.1
17g	> 150	8	3.2	30	15	< 5	28.8
17h	68	14	2.1	10.4	35.5	< 5	37.8

tive to the other prodrugs but are rapidly hydrolyzed by liver homogenate. Based on these promising *in vitro* results, prodrugs **17a–h** were evaluated for their pharmacokinetic properties in dogs, and all demonstrated good oral bioavailability (20–38%).[54] No metabolites were found other than PMPA. Under similar conditions, PMPA has 20% oral bioavailability in dogs.

Based on its stability, solubility and improved oral bioavailability over PMPA, bis(POC)PMPA **17f** was selected for further development.

VI. OTHER NUCLEOSIDE PHOSPHONATE PRODRUGS

In a closely related series of experiments, another methylene phosphonate drug, 9-[2-(phosphonomethoxy)ethoxy]adenine (**18**, BRL 47923), was derivatized to prepare a variety of possible prodrugs.[55]

BRL 47923 (**18**), an oxygen homolog of PMEA, is active against a number of retroviruses, including HIV, SIV, and FIV.[56] It suffers from low oral bioavailability similarly to other methylene phosphonate drugs. For example, when administered p.o. to mice only 2% of the drug was detected in the blood.[57]

A battery of candidate prodrugs were prepared (Figure 8) and administered to mice.[55] As with adefovir, the more lipophilic alkyl diesters **18a–d** were absorbed from the digestive tract but were converted to the parent drug only to a minor extent (10% for *i*-Pr, **18c** being the best). Similarly, the dibenzyl derivatives were absorbed, although to a lesser degree than the dialkyl esters, but their conversions were even worse (8% for **18e**). The quinone methide precursors[58] 4-isobutyryloxybenzyl **18f** and 4-acetoxybenzyl **18g** were also not useful, with bioavailabilities of 1 and 8%, respectively. Two diamidates, the bis-(diethylamino) **18h** and the bis-(glycine methyl ester)**18i**, were made, but neither gave detectable BRL 47923 in the blood.

Acyloxymethyl promoieties were as useful with BRL 47923 (**18**) as they were with adefovir. The bispivaloyloxymethyl derivative **18j** was 30% bioavailable and cleanly converted to the parent drug in the blood. The bispivaloyloxyethyl derivative **18k** was even more bioavailable, at 74%, again without observed intermediates or alternative metabolites. Despite this finding, the fact that the bispivaloyl-

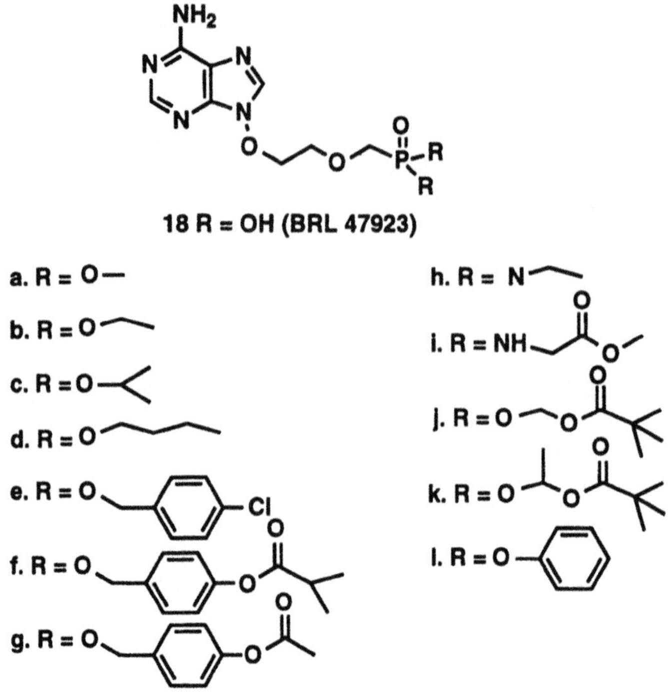

18 R = OH (BRL 47923)

a. R = O—

b. R = O⌒

c. R = O—<

d. R = O⌒⌒⌒

e. R = O...⟨C₆H₄⟩—Cl

f. R = O...⟨C₆H₄⟩—O—C(=O)—CH(CH₃)₂

g. R = O...⟨C₆H₄⟩—O—C(=O)—CH₃

h. R = N⌒⌒

i. R = NH—⌒—C(=O)—O—

j. R = O⌒O—C(=O)—C(CH₃)₃

k. R = O—CH(—O—C(=O)—C(CH₃)₃)—CH₃

l. R = O—⟨C₆H₅⟩

Figure 8.

oxyethyl derivative **18k** is a diastereomeric mixture was regarded as sufficient grounds to disqualify that prodrug from further consideration. Other acyloxyalkyl derivatives were made and tested, including a series of mixed esters with one pivaloyloxymethyl group and one alkyl group. Some of these were well absorbed, but none were converted all the way to the parent drug.

A series of aryl esters was also prepared. Bioavailability did not vary in any obvious way with substitution, although the increase in oral bioavailability of the diphenyl ester **18l** from 26 to 50% on conversion to the hydrochloride salt suggests that solubility and initial transport may be important. The diphenyl ester **18l** was also rather stable chemically and in mouse duodenal contents, so it was chosen to be the candidate prodrug. In contrast to the adefovir case, other metabolites were not observed on administration of the diphenyl ester.

VII. PRECLINICAL EVALUATION OF ADEFOVIR DIPIVOXIL AND BIS(POC)PMPA

Prodrugs of phosphonates are often transported into cells more efficiently than the parent drugs. This can yield much higher *in vitro* potencies which may or may not extrapolate into the *in vivo* situation. Adefovir dipivoxil and Bis(POC)PMPA were found to exhibit 9-23-fold greater anti-HIV activity. Metabolism studies with tritiated adefovir dipivoxil and tritiated PMEA showed > 100-fold increase in the cellular uptake of the prodrug derivative relative to PMEA and formation of active intracellular diphosphorylated metabolite (PMEApp).[53b] In another study PMEA was compared to adefovir dipivoxil and diphenyl PMEA.[49] Adefovir dipivoxil was at least 20 times more potent than PMEA against HIV, and 250 times more active against MSV (Moloney murine sarcoma virus). Diphenyl PMEA was also more potent than PMEA, but the effects were not as dramatic. In some cell lines relative toxicity increased with potency, so the selectivity index did not improve, whereas in others some gains were noted. Both adefovir dipivoxil and bis(POC)PMPA were evaluated in several mouse retroviral models. Adefovir dipivoxil was as effective as equimolar doses of subcutaneous adefovir against MSV[59] and FLV (Friend leukemia virus). Oral adefovir and the diphenyl ester **13d** were much less effective, correlating with their lower bioavailabilities. The apparent first-pass metabolism of adefovir dipivoxil in the liver may also lead to higher levels of the drug in the liver than are found with parenteral adefovir. This may have implications for the potential therapy of HBV, against which adefovir is active. Similarly, orally administered bis(POC)PMPA showed comparable efficacy to parenterally administered PMPA in MSV infected mice[60]

The toxicities[61] of adefovir and adefovir dipivoxil were investigated in a 13-week repeated dose study in monkeys. The monkeys received adefovir intravenously once daily at doses of 0, 3 or 20 mg/kg/day and adefovir dipivoxil via oral gavage once daily at doses of 0, 5 and 25 mg/kg/day for 13 weeks. Both adefovir and the oral prodrug have a common dose limiting target-organ toxicity (nephrotoxicity) and unique secondary toxicities that may reflect differences in pharmacokinetics and/or distribution profiles. For example, adefovir is associated with skin toxicity not observed with the prodrug and, conversely, the prodrug is associated with mild reversible hepatic toxicity not observed with the parent.

In single dose tissue distribution studies utilizing ^{14}C-labeled material,[61] oral administration of ^{14}C-adefovir dipivoxil resulted in low (< 0.1% of total dose) but measurable radioactivity in GI tract, Kidney and skeletal muscle tissues. No significant metabolites were observed. Adefovir was found to be excreted unchanged through the urine whereas orally administered adefovir dipivoxil was eliminated both in urine and feces.

VIII. CLINICAL STUDIES OF ADEFOVIR DIPIVOXIL AND BIS(POC)PMPA

Adefovir dipivoxil has been evaluated in three clinical studies. In a Phase I study, the bioavailability of adefovir dipivoxil when administered to fasting and fed patients was 30 and 41%, respectively, a statistically significant difference.[31] Also, pharmacokinetic analysis showed the exposure was dose proportional and, after 14 consecutive days of dosing, there was no evidence for drug accumulation.

In a subsequent Phase I/II study,[62] adefovir dipivoxil was administered at doses of 125 or 250 mg to HIV-infected patients in a double-blind, placebo-controlled manner in which patients were randomized to receive either adefovir dipivoxil or placebo. After 6 weeks of blinded treatment, the mean CD4 T-cell count had increased 46 and 15 cells/mm^3 in the 125 and 250 mg treatment groups, respectively, compared to a decline of 41 cells/mm^3 in the placebo; this result was statistically significant when comparing the placebo and 125 mg groups, p = 0.013. The increased CD4 cell count was maintained through week 12. At week 6, patients receiving placebo had no change from baseline in HIV-1 RNA levels. In contrast, patients receiving 125 mg daily of adefovir dipivoxil had a median decline of 0.5 log copies/ml (p = 0.002 compared to placebo, Wilcoxon rank sum test). This median 0.5 log decline remained stable through week 12. Similar results were obtained when all available timepoints were analyzed. Patients receiving 250 mg/day of adefovir dipivoxil had a median decline from baseline of 0.4 log at week 6 (p = 0.01 compared to baseline, Wilcoxon rank sum test). This decline also remained stable through week 12. Thus, adefovir dipivoxil administration is associated with significant and durable anti-HIV activity as measured by CD4 T-cell counts and viral load.

The most common adverse events associated with adefovir dipivoxil administration were gastrointestinal (primarily nausea and diarrhea). These events were mild to moderate, increased at the higher doses and were reversible. The 125 mg dose was well tolerated, without evidence of significant drug-related toxicity and was chosen for subsequent Phase II/III studies. Adefovir dipivoxil administration resulted in a drop in serum carnitine levels. While no patient developed symptoms of carnitine deficiency, and the clinical significance of reduced serum carnitine levels is unclear, adefovir dipivoxil is currently being provided with L-carnitine supplements.

The pharmacokinetics, safety and efficacy of orally administered bis(POC)PMPA was assessed in a Phase I/II dose escalation study.[63] Bis(POC)PMPA was administered to HIV infected adults in a double blind, placebo controlled manner at 75, 150, and 300 mg once daily doses. Oral bioavailability was 27% in the fasted state and 41% in the fed state and systemic exposure was dose proportional. The median $\log_{10}$ decreases in plasma HIV RNA levels from baseline after 28 days of dosing were 0.06, 0.32, 0.44, and 1.22 in the placebo, 75, 150, and 300 mg cohorts, respectively. The most common adverse events were reversible creatine kinase elevations.

Based on these promising results, bis(POC)PMPA is being evaluated in a phase II study.

IX. CONCLUSIONS

The considerable metabolic and therapeutic advantages of acyclic nucleotide drugs can be extended to oral administration forms by a prodrug strategy. The oral bioavailabilities of nucleotides from prodrugs can in some cases exceed 40%, more than adequate for practical clinical application. The approach to finding successful prodrug candidates is still substantially empirical, although the experience base continues to grow. *In vitro* chemical experiments are straightforwardly useful; chemical stability and solubility are unambiguously required. *In vitro* biochemical assays are interesting in retrospect, but not necessarily good predictors. Both false positive and false negative predictions have been found, and there is not yet a valid substitute for animal experimentation.

REFERENCES

1. Holy, A. In: *Advances in Antiviral Drug Design*; De Clercq, E., Ed.; JAI Press: Greenwich, CT, 1993, Vol. 1, pp 179–231.
2. O'Brien, J. J.; Campoli-Richards, D. M. *Drugs* **1989**, *37*, 233–309.
3. Morton, P.; Thompson, A.N. *NZ Med. Journal* **1989**, *102*, 93–95.
4. Perry, C. M.; Wagstaff, A. J. *Drugs* **1995**, *50*, 396–415.
5. Faulds, D.; Heel, R. C. *Drugs* **1990**, *39*, 597–638.
6. Crumpacker, C. S. *New Eng. J. Med.* **1996**, *335*, 721–729.
7. Langtry, H. D.; Campoli-Richards, D. M. *Drugs* **1989**, *37*, 408–450.
8. Coates, J. A. V.; Cammack, N. S.; Jenkinson, H. J.; Mutton, I. M.; Pearson, B. A.; Storer, R.; Cameron, J. M.; Penn, C. R. *Antimicrob. Chem. Chemother.* **1992**, *36*, 202–205.
9. Mitsuya, H.; Border, S. *Proc. Natl. Acad. Sci.* USA **1986**, *83*, 1911–1915.
10. Faulds, D.; Brogden, R. N. *Drugs* **1992**, *44*, 94–116.
11. Hitchcock, M. J. M. *Antiviral Chem. Chemother.* **1991**, *2*, 125–132.
12. De Clercq, E. *Rev. Med. Virol.* **1993**, *3*, 85–96.
13. Arends, S.; Van Halteren, E.; Kamp, W. *Pharmacy World & Science* **1996**, *18*, 30–34.
14. Pauwels, R.; Balzarini, J.; Scols, D.; Baba, M.; Desmyter, J.; Rosenberg, I.; Holy, A.; De Clercq, E. *Antimicrob. Agents Chemother.* **1988**, *32*, 1025–1030.
15. (a) Tsai, C. C.; Follis, K. E.; Sabo, A.; Beck, T. W.; Grant, R. F.; Bischofberger, N.; Benveniste, R. E.; Black, R. *Science* **1995**, *270*, 1197–1199. (b) Tsai, C. C.; Follis, K. E.; Beck, T. W.; Sabo, A.; Bischofberger, N.; Dailey, P. J. *Aids Res. Human Retroviruses* **1997**, *13*, 707–712.
16. Barditch-Crovo, P.; Deeks, S.; Kahn, J.; Redpath, I.; Smith, A.; Hwang, F.; Hellman, N.; Cundy, K. C.; Roony, J.; Lietman, P. *10th ICAR, Atlanta, Georgia.* 1997, April 6–11 (oral presentation).
17. Elion, G. B. *Am. J. Med.* **1982**, *73*, 7–13.
18. Elion, G. B. *J. Antimicrob. Chemother.* **1983**, *12 (Suppl. B.)*, 9–17.
19. Miller, W. H.; Miller, R. L. *J. Biol Chem.* **1980**, *255*, 7204–7207.
20. Keller, P. M.; Fyfe, J. A.; Beauchamp, L.; Lubbers, C. M.; Furman, P. A.; Schaeffer, H. J.; Elion, G. B. *Biochem. Pharmacol.* **1981**, *30*, 3071–3077.
21. Smee, D. F.; Aurelian, L.; Reddy, M. P.; Miller, P. S.; Tso, P. O. P. *Proc. Natl. Acad. Sci. USA* **1986**, *83*, 2787–2791.
22. Martin, J. C.; Dvorak, C. A.; Smee, D. F.; Mattews, T. R.; Verheyden, J. P. H. *J. Med. Chem.* **1983**, *26*, 759–761.
23. Littler, E.; Stuart, A. D.; Chee, M. S. *Nature* **1992**, *358*, 160–162.
24. Biron, K. K.; Stanat, S. C.; Sorrell, J. B.; Fyfe, J. A.; Keller, P. M.; Lambe, C. A.; Nelson, D. J. *Proc. Natl. Acad. Sci. USA* **1985**, *82*, 2473–2477.

25. Chou, S.; Guentzel, S.; Michels, K. R.; Miner, R. C.; Drew, W. L. *J. Infect. Dis.* **1995**, *172*, 239–42.
26. Sullivan, V.; Biron, K. K.; Talarico, C.; Stanat, S.; Davis, M.; Pozzi, L. M.; Coen, D. M. *Antimicrob. Agents Chemother.* **1993**, *37*, 19–25.
27. Gao, W.; Shirasaka, T.; Johns, D.; Broder, S.; Mitsuya, H. *J. Clin. Invest.* **1993**, *91*, 2326–2333.
28. (a) Landt, M.; Everard, R. A.; Butler, L. G. *Biochemistry* **1980**, *19*, 138–143. (b) Kelly, S. J.; Dardinger, D. E.; Butler, L. G. *Biochemistry* **1975**, *14*, 4983–88. c. Kelly, S. J., Butler, L. G. *Biochemistry* **1977**, *16*, 1102–1104.
29. Cohen, S. S.; Plunkett, W. *Ann. N.Y. Acad. Sci.* **1975**, *255*, 269–286.
30. (a) Starrett, J. E.; Mansuri, M. M.; Martin, J. C.; Tortolani, D. R.; Bronson, J. J. *Europ. Patent* **1992**, 481, 214, A1. (b) Cundy, K. C.; Lee, W. A. *Antimicrob. Agents Chemother.* **1994**, *38*, 365–368.
31. Cundy, K. C.; Barditch-Crovo, P; Walker, R. E.; Collier, A. C.; Ebeling, D.; Toole, J.; Jaffe, H. S. *Antimicrob. Agents Chemother.* **1995**, *39*, 2401–2405.
32. Fleit, H.; Simpson, R. J.; Webster, A. D. B.; Peters, T. J. *J. Biol. Chem.*, **1975**, *250*, 8889–8892.
33. Cihlar, T.; Rosenberg, I.; Votruba, I.; Holy, A. *Antimicrob. Agents Chemother.* **1995**, *39*, 117–124.
34. Connelly, M. C.; Robbins, B. L.; Fridland, A. *Biochem. Pharmacol.* **1993**, *46*, 1053–1057.
35. Balzarini, J.; Hao, Z.; Herdewijn, P.; Johns, D. G.; De Clercq, E. *Proc. Natl. Acad. Sci. USA* **1991**, *88*, 1499–1503.
36. Hitchcock, M. J. M.; Jaffe, H. S.; Martin, J. C.; Stagg, R. J. *Antiviral Chem. and Chemother.* **1996**, *7*, 115–127.
37. Puech, F.; Gosselin, G.; Lefebvre, I.; Pompon, A.; Aubertin, A.-M.; Dirn, A.; Imbach, J.-L. *Antiviral Res.* **1993** , *22*, 155–174.
38. Jones, R. J.; Bischofberger, N. *Antiviral Res.* **1995**, *27*, 1–17.
39. Krise, J. P.; Stella, V. J. *Adv. Drug Deliv. Rev.* **1996**, *19*, 287–310.
40. Hilgers, A. R.; Conradi, R. A.; Burton, P. S. *Pharm. Res.* **1990**, *7*, 902–910.
41. Hidalgo, I. J.; Hillgren, G. M.; Grass, G. M.; Borchardt, R. T. *Pharm. Res.* **1991**, *8*, 222–227.
42. Shaw, J.-P.; Cundy, K. C. *Pharm. Res.* **1993**, *10 (Suppl.)*, S294.
43. Mukaiyama, T.; Hashimoto, M. *J. Am. Chem. Soc.* **1972**, *94*, 8528–8532.
44. Starrett Jr., J. E.; Tortolani, D. R.; Russell, J.; Hitchhock, M. J. M.; Whiterock, V.; Martin, J. C.; Mansuri, M. M. *J. Med. Chem.* **1994**, *37*, 1857–1864.
45. Starrett Jr., J. E.; Tortolani, D. R.; Hitchhock, M. J. M.; Martin, J. C.; Mansuri, M. M. *Antiviral Res.* **1992**, *19*, 267–273.
46. Alexander, P.; Holy, A.; Masojidkova, M. *Coll. Czech. Chem. Comm.* **1994**, *59*, 1853–1869.
47. Bundgaard, H. *Drugs Future* **1991**, *16*, 443–458.
48. (a) Shaw, J.-P.; Jones, R. J.; Arimilli, M. N.; Louie, M.S .; Lee, W. A.; Cundy, K. C. *Pharm. Res. Suppl.* **1994**, *11*, S402. (b) Shaw, J.-P.; Louie, M. S.;

Krishnamurthy, V. V.; Arimilli, M. N.; Jones, R. J.; Bidgood, A.; Lee, W. A.; Cundy, K. C. *Drug Metab. Dispos.* **1997**, *25,* 361–366.

49. Naesens, L.; Balzarini, J.; Bischofberger, N.; De Clercq, E. *Antimicrob. Agents Chemother.* **1996**, *40,* 22–28.

50. (a) Jones, R. J.; Arimilli, M. N.; Lin, K.-Y.; Louie, M. S.; Mcgee, L. R.; Shaw, J.-P.; Burman, D.; Lee, M. L.; Kennedy, J. A.; Prisbe, E. J.; Bischofberger, N.; Lee, W. A.; Cundy, K. C. *XII Intl. Roundtable on Nucleosides and Nucleotides, La Jolla, CA.* 1996, p 11 (oral presentation). (b) Cundy, K. C.; Fishback, J. A.; Shaw, J.-P; Lee, M. L.; Soike, K. F.; Visor, G. C.; Lee, W. A. *Pharm. Res.* **1994**, *11,* 839–843. (c) Cundy, K. C.; Shaw, J.-P. Personal communication.

51. Abrahamsson, K.; Ericksson, B.O.; Holme, E.; Jodal, U.; Lindstedt, S.; Nordin, I. *Biochem. Med. Metabol. Biol.* **1994**, *52,* 18–21.

52. Ikeda, S.; Sakamoto, F.; Kondo, H.; Moriyama, N.; Tsukamoto, G. *Chem. Pharm. Bull.* **1984**, *32,* 4316–4322.

53. (a) Arimilli, M. N.; Kim, C.U.; Dougherty, J.; Mulato, A.; Oliyai, R.; Shaw, J.-P.; Cundy, K. C.; Bischofberger, N. *Antiviral Chem. Chemother.* **1997**, *8,* 557–564. (b) Srinivas, R.V.; Robbins, B. L.; Fridland, A. *Biochem. Pharmacol.* **1993**, *46,* 1053–1057.

54. Shaw, J.-P.; Sueoka, C. M.; Oliyai, R.; Lee, W. A.; Arimilli, M. N.; Kim, C. U.; Cundy, K. C. *Pharm. Res.* **1997**, *14,* 1823–1828.

55. Serafinowska, H. T.; Ashton, R. J.; Bailey, S.; Harnden, M. R.; Jackson, S. M.; Sutton, D. *J. Med. Chem.* **1995**, *38,* 1372–1379.

56. Perkins, R. M.; Immelmann, A.; Elphick, L.; Duckworth, D. M.; Harnden, M. R.; Kenig, M. D.; Planterose, D. N.; Brown, A. G. *Antiviral Res.* **1992**, *17* (Suppl I) 59.

57. Perkins, R. M.; Barney, S.; Wittrock, R.; Clark, P. H.; Levin, R.; Lambert, D. M.; Petteway, S. R.; Serafinowska, H. T.; Bailey, S.; Jackson, S. M.; Harnden, M. R.; Ashton, R. J.; Sutton, D.; Harvey, J. J.; Brown, A. G. *Antiviral Res.* **1993**, *20,* Suppl. I, 84.

58. Mitchell, A. G.; Thomson, W.; Nicholls, D.; Irwin, W. J.; Freeman, S. *J. Chem. Soc., Perkin Trans. I* **1992**, 2345–2353.

59. Naesens, L.; Balzarini, J.; Bischofberger, N.; De Clercq, E. *Antimicrob. Agents Chemother.* **1996**, *40,* 22–28.

60. Bischofberger, N.; Naesens, D.; De Clercq, E.; Fridland, R. V.; Srinivas, R. V.; Robbins, B. L.; Arimilli, M. N.; Cundy, K. C.; Kim, C. U.; Lacy, S.; Lee, W. A.; Shaw, J.-P. *4th Conference on Retroviruses and Oppurtunistic Infections Washington, DC,* 1997, p 214.

61. Lacy, S.; Cundy, K. C.; Hitchcock, M. J. M. *National Society of Toxicology Meetings, Baltimore, MD.,* 1995, Vol.15, p 179.

62. Deeks, S. G.; Collier, A.; Lalezari, J.; Pavia, A.; Rodrigue, D.; Drew, W. L.; Toole, J.; Jaffe, H. S.; Mulato, A. S.; Lamy, P. D.; Li, W.; Cherrington, J. M.; Hellmann, N.; Kahn, J. *J. Infect. Diseases* **1997**, *176,* 1517–1523.

63. Deeks, S. G.; Barditch-Crovo, P.; Lietman, P. S.; Collier, A.; Safrin, S.; Coleman, R.; Cundy, K. C.; Kahn, J. O. *5th Conference on Retroviruses and Oppurtunistic Infections*, 1998.

HEPT: FROM AN INVESTIGATION OF LITHIATION OF NUCLEOSIDES TOWARDS A RATIONAL DESIGN OF NON-NUCLEOSIDE REVERSE TRANSCRIPTASE INHIBITORS OF HIV-1

Hiromichi Tanaka, Hiroyuki Hayakawa,
Kazuhiro Haraguchi, Tadashi Miyasaka,
Richard T. Walker, E. De Clercq,
Masanori Baba, David K. Stammers, and
David I. Stuart

Advances in Antiviral Drug Design
Volume 3, pages 93–144.
Copyright © 1999 by JAI Press Inc.
All rights of reproduction in any form reserved.
ISBN: 0-7623-0201-1

I. INTRODUCTION

Lithiation chemistry of aromatic compounds started with a halogen-lithium exchange reaction,[1,2] reflecting the historical finding that an aryllithium can be generated by treatment of an aryl halide with alkyllithium.[3] Soon afterwards, functional group-directed exchange of an aromatic hydrogen atom with lithium was discovered in two independent studies of *ortho* lithiation of anisole.[4–6] A wide variety of heteroatom-containing functional groups have been found to direct lithiation to their *ortho* position.[7] Since the lithiated species react with a wide range of carbon electrophiles, this strategy furnished a regio-defined C–C bond formation reaction leading to polysubstituted aromatic compounds.

Besides the *ortho* lithiation, numerous experimental results on the lithiation of heteroaromatic compounds have revealed an additional mechanism, *alpha* lithiation, wherein deprotonation occurs at the sp^2-hybridized carbon atom *alpha* to the heteroatom in the ring (Scheme 1).[8] These mechanistic understandings allow rational synthetic planning of target molecules. As a result, the lithiation chemistry has now gained a firm place, both in aromatic and heteroaromatic systems, as a highly general and efficient alternative to the classical C–C bond forming reactions.

In the field of nucleoside chemistry, construction of C–C bonds in the base moiety had been achieved in very few cases and it had mostly

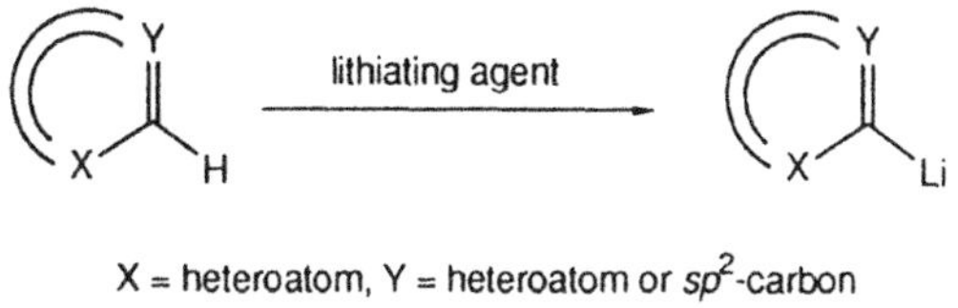

X = heteroatom, Y = heteroatom or sp^2-carbon

Scheme 1. The *alpha* lithiation of heteroaromatic compounds.

relied on the susceptibility of highly polarized purine and pyrimidine bases to nucleophilic reactions.[9,10] The pioneering work regarding lithiation of nucleosides appeared in 1959, which dealt with the halogen-lithium exchange reaction of 5-bromo-2′-deoxyuridine with butyllithum.[11] After quenching the resulting 5-lithiated species with [14C]-labeled MeI, [14C-Me]thymidine was obtained in 10% yield. Later reinvestigation by employing trimethylsilyl protection showed concomitant formation of the 6-lithiated species in this reaction.[12–14]

The first report of a hydrogen-lithium exchange reaction of nucleosides appeared in 1973.[15] In this instance, 2′,3′,5′-tris-*O*-trimethylsilyl (TMS) uridine was deprotonated with butyllithium. As shown in Scheme 2, however, the result was not satisfactory in respect of either the extent of lithiation (42%) or the regioselectivity (C5 vs. C6), which varies depending on the electrophile used (D_2O or $^{14}CH_3I$).

Presumably due to these rather discouraging results, together with a lack of a suitable sugar-protecting group, the synthetic application

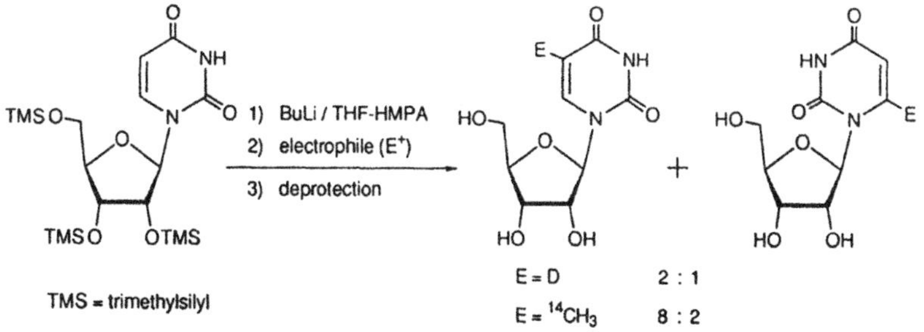

TMS = trimethylsilyl

E = D   2 : 1

E = $^{14}CH_3$ 8 : 2

Scheme 2. Hydrogen-lithium exchange reaction of 2′,3′,5′-tris-*O*-(trimethylsilyl)uridine with butyllithium.

of lithiation chemistry in this field had long been limited to the preparation of radiolabeled derivatives. The focus of this present article will be given first to the *alpha* lithiation reaction at the base moiety of purine, imidazole, and pyrimidine nucleosides for the synthesis of their analogs, showing principally our own results, and then to an anti-HIV-1 (Human Immunodeficiency Virus Type 1) lead compound HEPT which resulted from our lithiation studies.

II. LITHIATION OF PURINE NUCLEOSIDES AT THE 8- OR 2-POSITIONS

A brief report published in 1979 by Barton et al. that alkylation at the 8-position of N^6-methylated 2′,3′-O-isopropylideneadenosine can be achieved by way of its lithio intermediate, constitutes the first example of lithiation of purine nucleosides (Scheme 3).[16] In this report, 2′,3′-O-isopropylidene-N^6-methyladenosine (**1**) was reacted with butyl-lithium in THF at −78 °C followed by MeI. The product **2** was obtained in 35% yield, along with a small amount of the 5′-O-methylated by-product **4** (7%). Similar methylation and ethylation of the fully

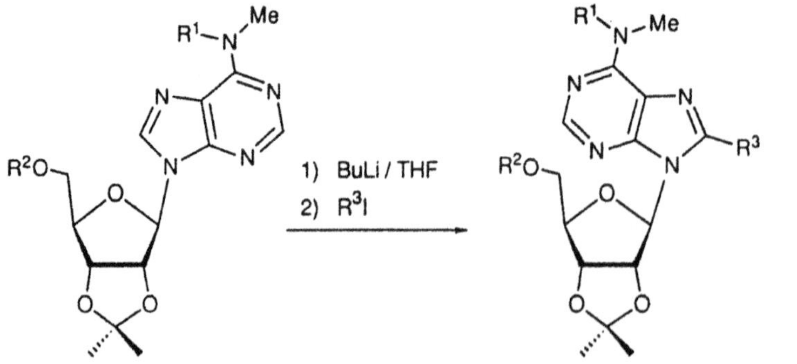

1 $R^1 = R^2 = H$
3 $R^1 = R^2 = Me$

2 $R^1 = Me, R^2 = H, R^3 = Me$ (35%)
4 $R^1 = R^2 = R^3 = Me$ (88%)
5 $R^1 = R^2 = Me, R^3 = Et$ (51%)

Scheme 3.   The first example of C8-alkylation of purine nucleosides *via* the lithio intermediate.

protected derivative **3** gave compounds **4** and **5**, respectively, in higher yields.

We thought that the higher yields observed by using compound **3** could be attributable to improved solubility of the lithiated species. To eliminate the possibility of undesired N^6-alkylation, we selected 6-chloro-9-(2,3-*O*-isopropylidene-β-D-ribofuranosyl)purine (**6**) as a substrate for lithiation.[17] An additional advantage of the use of **6** is the anticipated susceptibility of the resulting products to nucleophilic substitution at the 6-position which would furnish a range of purine nucleoside analogs.

Lithiation of **6** with butyllithium (in THF at below −70 °C) followed by quenching with CD₃OD, however, gave a complex mixture of products, from which a small amount of the 8-butyl-6-chloro-9-(2,3-*O*-isopropylidene-β-D-ribofuranosyl)purine was isolated. This result suggests that partial halogen-lithium exchange had taken place to form butyl chloride, which then reacted with the 8-lithiated species.[18] In contrast to this, when LDA (lithium diisopropylamide) was used instead of butyllithium, no by-product was formed and **7** was obtained in 93% isolated yield with 81% of deuterium incorporation at the 8-position (Scheme 4).

Carbon electrophiles such as MeI, PhCHO, EtCHO, Ph₂CO and Et₂CO were used to give compounds **8–12**. The formation of **8** and its low yield indicate that the lithio intermediate was consumed in deprotonating the initially introduced 8-methyl group. An enolizable ketone decreased yield of product as can be seen by comparison of the yields between **11** and **12**.[18] Conversion of 8-substituted products to other purine nucleosides is exemplified by the formation of **14** (NaSH/DMF) and **15** (NH₄OH/THF) from **13**, which was prepared by MnO₂-oxidation of **9**, as well as by that of **16** (H₂/Pd-C/Et₃N/aqueous EtOH) from **8**. Later, this LDA lithiation method was found to be applicable to naturally occurring purine nucleosides by employing *tert*-butyldimethylsilyl sugar protection.[19–21]

In a recent study, we found an alternative way of utilizing lithiation of 6-chloropurine nucleosides, this time for the synthesis of 2-substituted derivatives.[22] In accord with the above result of **6**, 9-[2,3,5-tris-*O*-(*tert*-butyldimethylsilyl)-β-D-ribofuranosyl]-6-chloropurine (**17**) undergoes exclusive lithiation at the 8-position to the extent of 89% upon treatment with LDA (in THF, below −70 °C). This was confirmed by

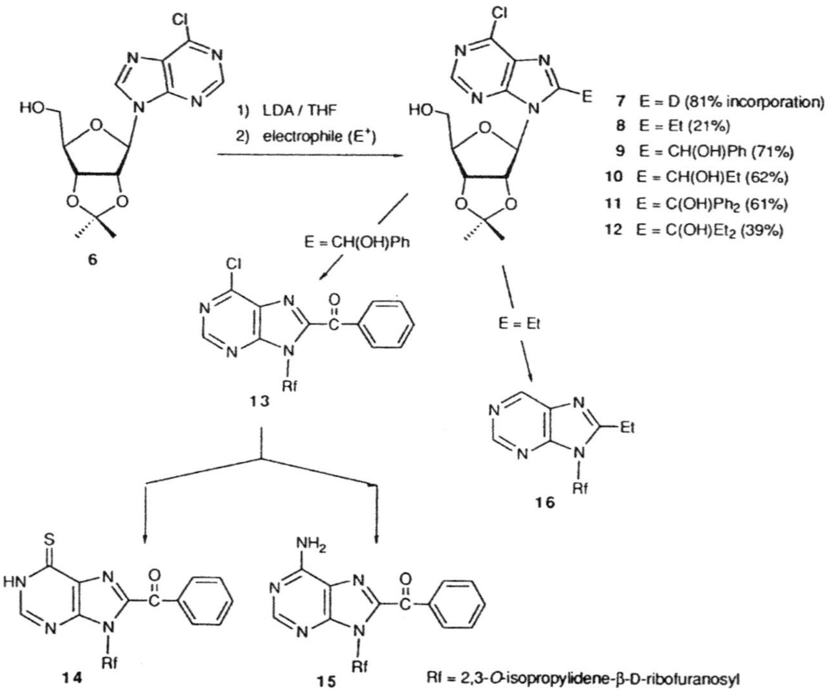

HO
Cl
6
1) LDA / THF
2) electrophile (E⁺)
E
7 E = D (81% incorporation)
8 E = Et (21%)
9 E = CH(OH)Ph (71%)
10 E = CH(OH)Et (62%)
11 E = C(OH)Ph2 (61%)
12 E = C(OH)Et2 (39%)
E = CH(OH)Ph
13
E = Et
Et
16
S
HN
14
NH2
15
Rf
Rf = 2,3-O-isopropylidene-β-D-ribofuranosyl

Scheme 4. LDA-lithiation of 6-chloro-9-(2,3-*O*-isopropylidene-β-D-ribofuranosyl)purine (**6**) and conversion of C8-substituted products to other types of purine nucleosides.

deuterium incorporation as well as by the preparation of the 8-iodo derivative **18** as shown in Scheme 5. However, when the same lithiated reaction mixture was quenched with TMSCl, five products **19–23** were obtained together with a considerable recovery (33%) of **17**. Apart from question why the 2-substituted compounds **19** and **21** were produced, it is apparent that **22** and **23** resulted from nucleophilic displacement with LDA. The fact that such undesired displacement was not observed during the formation of **18** suggests that the TMSCl was acting as a Lewis acid to coordinate with a nitrogen of the purine ring, for example with N1.

It can be readily anticipated that a more bulky lithium dialkylamide could eliminate such a pathway. When LTMP (lithium 2,2,6,6-tetra-methylpiperidide), which is more bulky and also more basic than LDA, was employed as a lithiating agent, only **19**, **20**, and **21** were formed, but again **17** was recovered in 40% yield. A proposed possible reaction mechanism in Scheme 6 came from the following experimental results: (1) no change was seen upon treating **21** with LTMP, showing that the 2-TMS group once introduced is stable under these lithiation conditions; (2) LTMP treatment of the 8-TMS derivative **20** gave **17**, **19**, and **21**; and (3) when the isolated 2,8-bis-TMS derivative **19** was reacted with the lithiated species of **17**, compound **21** was formed in 72% yield (based on the combined amounts of **17** plus **19**), suggesting that compound **21** resulted not only from **19**, through direct removal of the 8-TMS group, but also from **17**.

It appeared that addition of HMPA (hexamethylphosphoric triamide) in this reaction furnished the 2-TMS derivative **21** exclusively in quantitative yield, presumably by increasing silaphilicity of the lithiated species. An important feature of this reaction is that stannyl chloride can be used as an electrophile, instead of TMSCl, as shown in Scheme 5 by the formation of the 2-stannyl derivative **24**. This enabled further functionalization at the 2-position by replacement of the stannyl group.

Scheme 7 shows such functionalizations starting from either **24** or its adenosine counterpart **25** which was obtained by ammonolysis of **24**.

These compounds readily undergo halogenation by the use of iodine or *N*-halosuccinimides to give **26–31**. A high yield preparation of 2-fluoro derivatives (**32** and **33**) was also accomplished by using

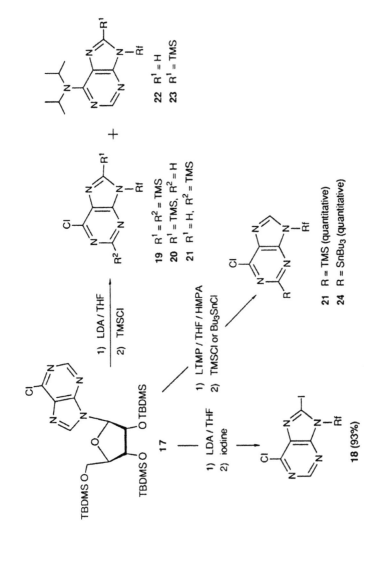

Rf = 2,3,5-tris-O-(tert-butyldimethylsilyl)-β-D-ribofuranosyl

Scheme 5. Lithiation of 9-[2,3,5-tris-O-(tert-butyldimethylsilyl)-β-D-ribofuranosyl]-6-chloropurine (**17**).

100

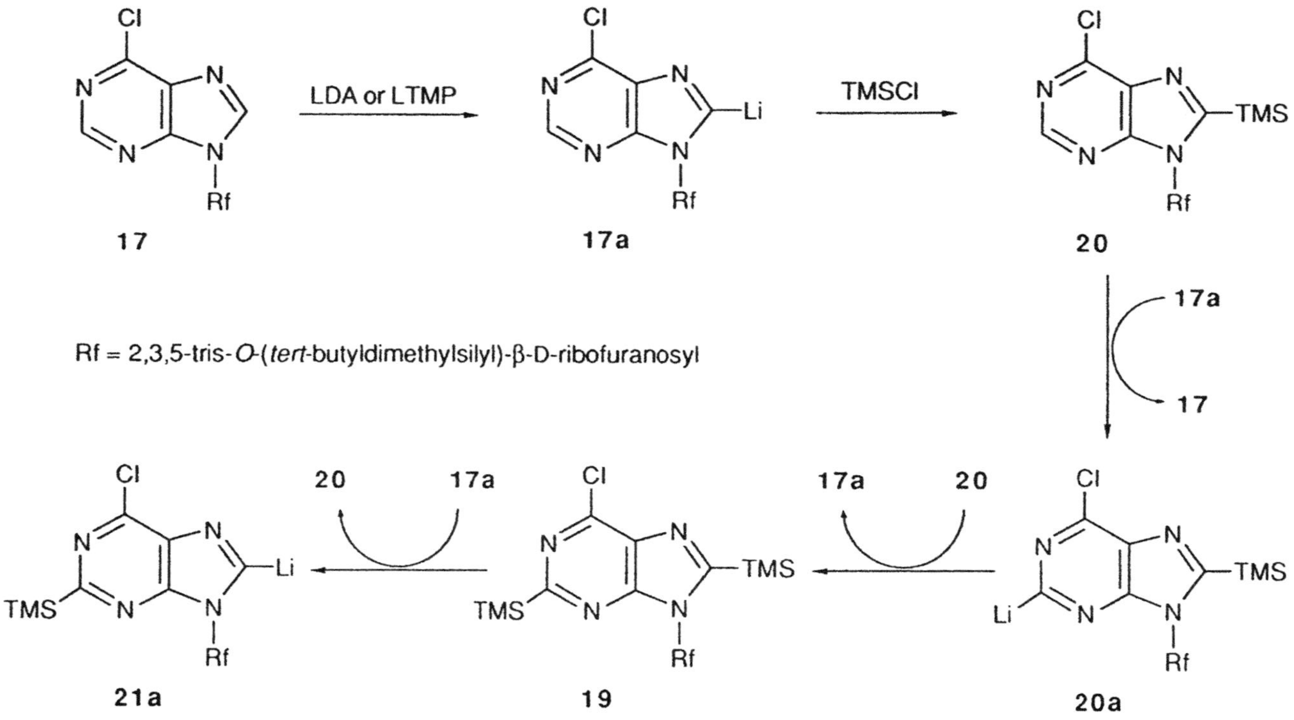

Scheme 6. A possible reaction mechanism for the formation of the 2-trimethylsilyl derivative **21** from compound **17**.

Rf = 2,3,5-tris-*O*-(*tert*-butyldimethylsilyl)-β-D-ribofuranosyl

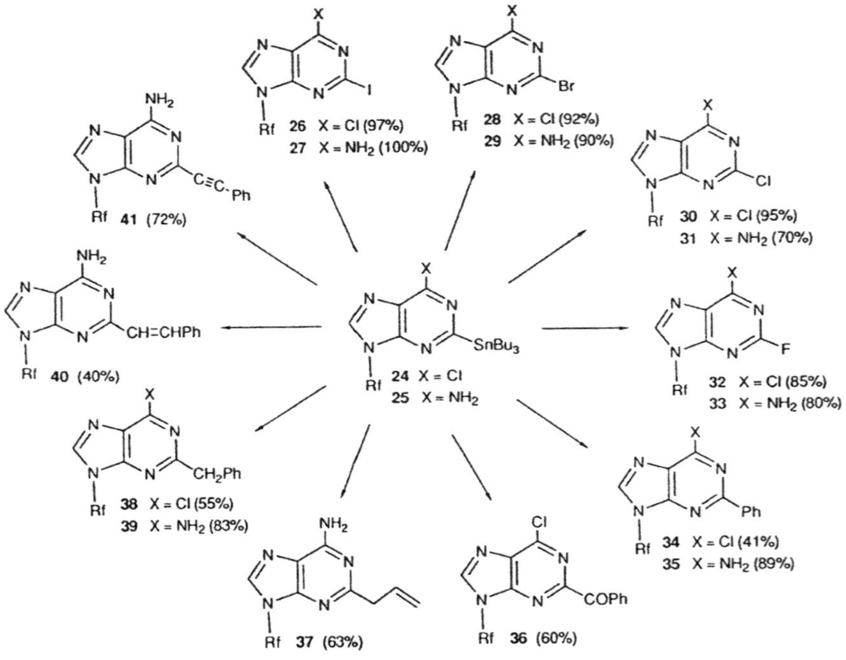
26 X = Cl (97%)
27 X = NH2 (100%)
28 X = Cl (92%)
29 X = NH2 (90%)
30 X = Cl (95%)
31 X = NH2 (70%)
32 X = Cl (85%)
33 X = NH2 (80%)
34 X = Cl (41%)
35 X = NH2 (89%)
36 (60%)
37 (63%)
38 X = Cl (55%)
39 X = NH2 (83%)
40 (40%)
41 (72%)
24 X = Cl
25 X = NH2
SnBu3
CH=CHPh
CH2Ph
COPh

Rf = 2,3,5-tris-*O*-(*tert*-butyldimethylsilyl)-β-D-ribofuranosyl

Scheme 7. Introduction of various functional groups to the 2-position of compounds **24** and **25**.

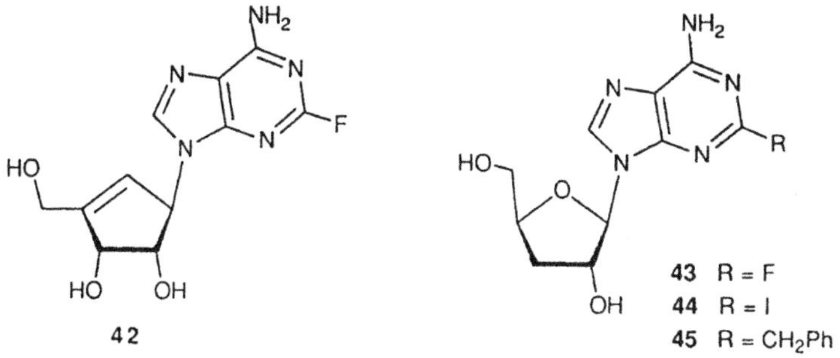

Figure 1. Some 2-substituted analogs of neplanocin A and cordycepin.

XeF$_2$ in the presence of AgOTf and 2,6-di-(*tert*-butyl)-4-methylpyridine.[23] For C–C bond formation, the Stille reaction,[24] palladium-catalyzed cross-coupling between organotin and RX, can be successfully applied to both **24** and **25** as shown by their conversion to **35–41**.[25]

It may deserve a mention that there have been a limited number of methods[26–29] for constructing C–C bonds at the 2-position of purine nucleosides. Moreover, except a homolytic methylation[26] and nucleophilic substitution with cyanide,[27] other methods involve the use of 6-chloro-2-iodopurine nucleosides, the preparation of which necessitates precursors having a 2-amino group for generating the purin-2-yl radical.[30] In contrast to this, the present method offers a certain advantage in that the synthesis of 2-substituted analogs of some adenine nucleoside antibiotics such as neplanocine A and cordycepin[31] can be accomplished through lithiation of their 6-chloro derivatives. This is illustrated in Figure 1 by the preparation of **42–45**.[32]

III. LITHIATION OF AN IMIDAZOLE NUCLEOSIDE AT THE 5-POSITION

The potential chemotherapeutic importance of 5-substituted imidazole nucleosides as analogs of naturally occurring 5-amino-1-β-D-ribofuranosylimidazole-4-carboxamide (AICAR, **46**) and bredinin (**47**),[33] has stimulated their synthesis. Apart from the classical conden-

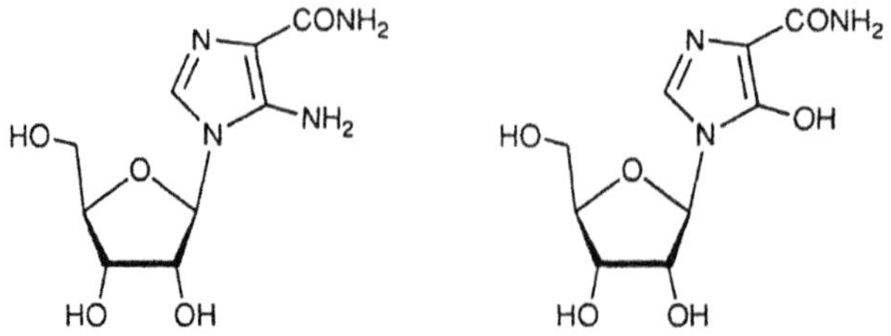

Figure 2. Structures of AICAR (46) and bredinin (47).

sation method between 5-substituted imidazole bases and protected sugar derivatives,[34] the Sandmeyer reaction of 5-amino-1-(2,3,5-tri-O-acetyl-β-D-ribofuranosyl)imidazole-4-carbonitrile, which gives 5-halogeno derivatives, had been the sole available method for their synthesis.[35,36] This fact motivated us to develop a new entry to the 5-substituted imidazole nucleosides based on lithiation chemistry.

It is well known that lithiation of 1-substituted imidazoles occurs at the carbon between nitrogen atoms.[37] Thus, to generate 5-lithiated species of an imidazole nucleoside, the accessibility of a derivative having a protecting group at the 2-position was crucial. As shown in Scheme 8, the 2-imidazolone nucleoside 48 was selected as the starting material for the following reasons: (1) this compound can be prepared from uridine in relatively large quantities;[38,39] and (2) the 2-chlorine atom of 49 is expected to be compatible with lithiation conditions and to be removable by hydrogenolysis.

The sugar hydroxyl groups of 49 were protected with 2',3'-O-methoxymethylidene and 5'-O-TBDMS groups to give the substrate 50 for the lithiation reaction. Upon lithiation of 50 with LDA, no chlorine-lithium exchange took place and, after quenching with electrophiles (MeI, PhCOCl, ClCOOMe, PhSSPh, and ClSiMe$_3$), various types of 5-substituted products (51–55) were obtained in good yields.[40] Hydrogenolysis of the 2-chlorine atom and successive deprotection are illustrated in Scheme 8 by the preparation of 56–58. As an application of this lithiation method, the synthesis of an antiviral 3-deazaguanosine (61)[41] was also carried out, which started with the

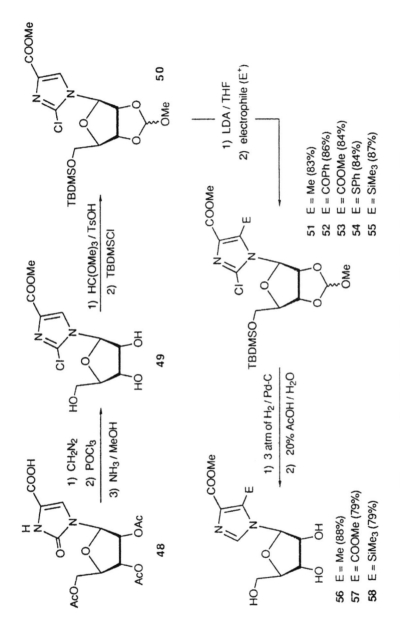

Scheme 8. Synthesis of 5-substituted imidazole nucleosides.

105

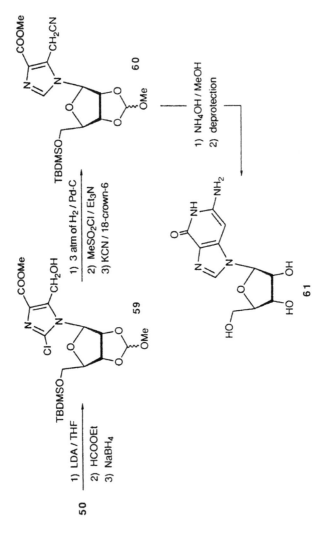

Scheme 9. Synthesis of 3-deazaguanosine based on LDA lithiation of compound **50**.

reaction between the lithiated species of **50** and ethyl formate as given in Scheme 9.[42] This LDA lithiation approach was also applicable to a substrate having a primary carboxamide, 2-chloro-1-(2,3-*O*-methoxy-methylidene-5-*O*-TBDMS-β-D-ribofuranosyl)imidazole-4-carboxamide, although the yields of 5-substituted products were uniformly lower.[43]

IV. LITHIATION OF URIDINE DERIVATIVES

A. General Approach to 6-Substituted Uridines

As mentioned in the introduction, lithiation of 2′,3′,5′-tris-*O*-(trimethylsilyl)uridine with butyllithium takes place at both the 5- and 6-positions (Scheme 2). In contrast, we found that 2′,3′-*O*-iso-propylideneuridine (**62**) undergoes lithiation-based alkylation with LDA exclusively at the 6-position. For example, treatment of **62** with LDA (in THF at below −70 °C) followed by MeI gave the 6-methyl-(**63**) and 6-ethyl- (**64**) derivatives (Scheme 10).[44] Although further alkylated product like **64** was inevitably formed,[45] it was interesting to see that, even by the use of excess LDA, no trace of any 5-alkylated product was detected in this alkylation. This result was initially explained by the possible participation of the 5′-oxygen atom in stabilizing the 6-lithiated species, but this explanation became untenable since regiospecific LDA lithiation was also observed in the case of 5′-deoxy-2′,3′-*O*-isopropylideneuridine.[46]

It is well known that application of the classical condensation method to 6-substituted pyrimidine bases almost always results in the predominant formation of N3-glycosylated products because of the steric hindrance of the 6-substituent.[47] Therefore, synthesis of 6-substituted pyrimidine nucleosides would be best carried out by the transformation of naturally occurring nucleosides. As the 6-position of the pyrimidine ring can be regarded to be the β-position of a kind of enone system, one would readily anticipate a nucleophilic addition-elimination mechanism, which is generalized in Scheme 11 for the case of uracil nucleosides.

An apparent severe limitation seen in this approach is that the "nucleophile (Nu)" has to be highly electron-withdrawing to make H-6 sufficiently acidic which is essential for regenerating the 5,6-double bond from the 5,6-dihydro intermediate. As a result, the most successful

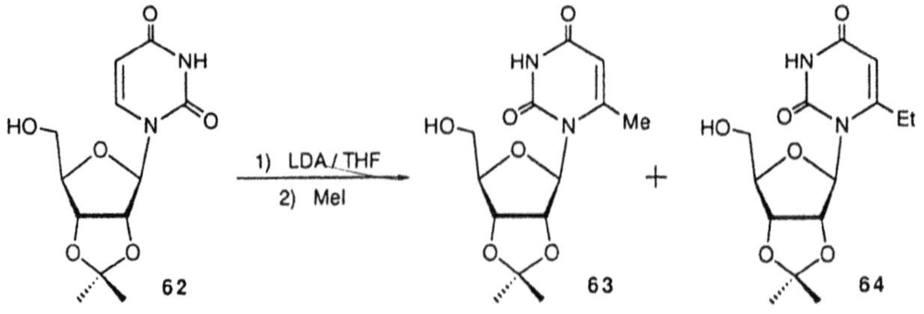

Scheme 10. LDA lithiation-based methylation of 2′,3′-O-iso-propylideneuridine (**62**).

example of this approach has been realized by using cyanide as a nucleophile in the reaction with 5-bromo derivatives of uridine, 2′-deoxyuridine, and cytidine.[48] The use of toluenethiolate as a nucleophile in such a reaction furnishes almost equal amounts of 5- and 6-substituted products.[49]

The lack of a general method for synthesizing 6-substituted pyrimidine nucleosides,[50] as well as the fact that organolithium intermediates react with a wide range of electrophiles under very mild conditions, led us to extend the above-mentioned regiospecific LDA lithiation of **62**. To make the 6-lithiated species sufficiently soluble, the 5′-O-methoxymethyl derivative (**65**, R = CH_2OMe) was prepared.[51] It was found later that the crystalline 5′-O-TBDMS derivative

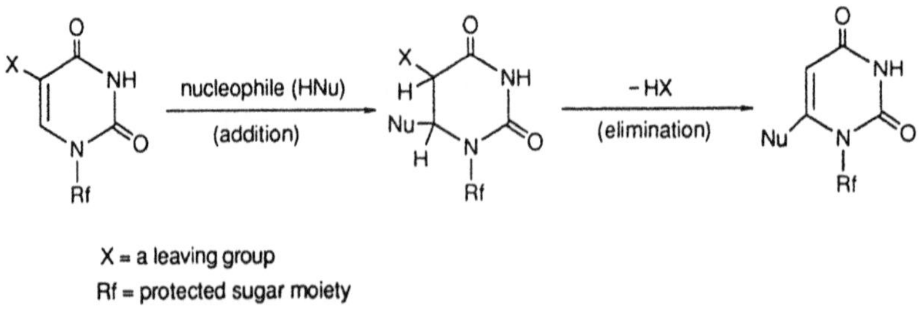

X = a leaving group
Rf = protected sugar moiety

Scheme 11. Nucleophilic addition-elimination pathway to 6-substituted uracil nucleosides.

(**65**, R = TBDMS) is a more convenient substrate because of its ease of preparation.[45] The extent (87–88%) of lithiation of **65** and its regiospecificity were confirmed based on ^{1}H NMR spectroscopic analysis of the 6-deuterated derivative.

As shown in Scheme 12, simply by reacting with electrophiles, a range of 6-substituted uridines can be synthesized.[51,52] As the 6-lithiated species derived from **65** is soluble in THF, alkylation can be carried out by an inverse addition technique, which avoids formation of further alkylated product.[53] Some of these compounds serve as starting materials for further transformations.[54,55] When the introduced 6-substituent has a good leaving ability, such as iodo and phenylthio, nucleophilic addition-elimination proceeds under very mild conditions.[56,57] This compensates for some limitations associated with the LDA lithiation approach.

One feature of 6-substituted pyrimidine nucleosides is their reverse glycosidic conformation to that found in naturally occurring nucleosides.[58] In spite of this fact, 6-phenylthio- (**66**, R = H, X = SPh) and 6-iodo- (**66**, R = H, X = I) uridines synthesized in our study were found to show activity against murine leukemia L5178Y cells in culture (Figure 3).[59]

To find more potent compounds, a series of their 5-substituted analogs was synthesized by applying the LDA-lithiation method to the corresponding 2′,3′-*O*-isopropylidene-5′-*O*-methoxymethyl derivatives. The activity of these analogs is listed in Figure 3. We initially considered the possibility that the activity would be a consequence of reaction of some biological nucleophiles with these compounds through the addition-elimination mechanism, but no apparent relationship was seen between the electronic effect of the 5-substituents and activity of the compounds.

During the above LDA lithiation of the 5-halogeno 2′,3′-*O*-isopropylidene-5′-*O*-methoxymethyluridines, by-products resulting from aryne formation, nucleophilic attack of the lithiating agent, or halogen-lithium exchange were hardly detected. However, lithiation of the 5-methyluridine derivative, where the electron-donating effect of the methyl group can participate in decreasing the acidity of H-6, results in a significantly reduced yield of the 6-substituted product (ca. 30%).

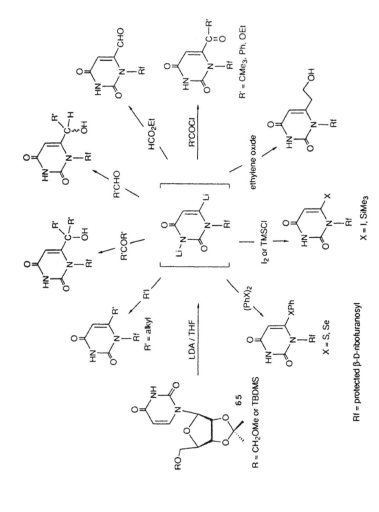

Scheme 12. Synthesis of various types of 6-substituted uridines based on regiospecific LDA lithiation of compound **65**.

110

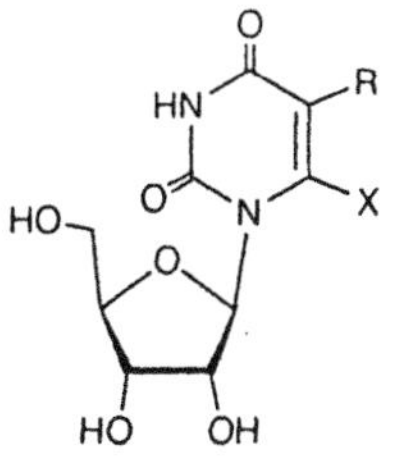

R	X	EC_{50} (μg / mL)[a]
H	SPh	70
Me	SPh	60
F	SPh	30
Cl	SPh	8.0
Br	SPh	20
I	SPh	40
H	I	8.0
Me	I	4.0
F	I	0.02
Cl	I	55
Br	I	52
I	I	55
5-fluorouracil		0.1

66

[a] Effective concentration of compound required to achieve 50% suppression of proliferation of mouse leukemia L5178Y cells in culture.

Figure 3. Structures and antileukemic activity of 6-phenylthio- and 6-iodouridine derivatives (**66**).

B. Lithiation of 6-Phenylthiouridine

The LDA lithiation method of 2′,3′-*O*-isopropylidene-5′-*O*-methoxymethyluridine (**65**) has afforded ready and high yield access to the 6-heteroatom-substituted uridines such as compounds **67–69**. Provided that their 5-lithiated species can be generated and the 6-substituent can be removed thereafter, it would lead to a new method for the synthesis of 5-substituted uridines.

The 6-iodo (**67**) and 6-trimethylsilyl (**68**) derivatives can be readily converted back to **65** in quantitative yield at room temperature under the conditions shown in Scheme 13. However the result of lithiation of these compounds was discouraging. Compound **67** was highly susceptible to halogen-lithium exchange to generate ca. 50% of the 6-lithiated species even with the use of LDA. Compound **68**, on the other hand, was recovered unchanged after treatment with several lithiating agents such as LDA, butyllithium, and LTMP.

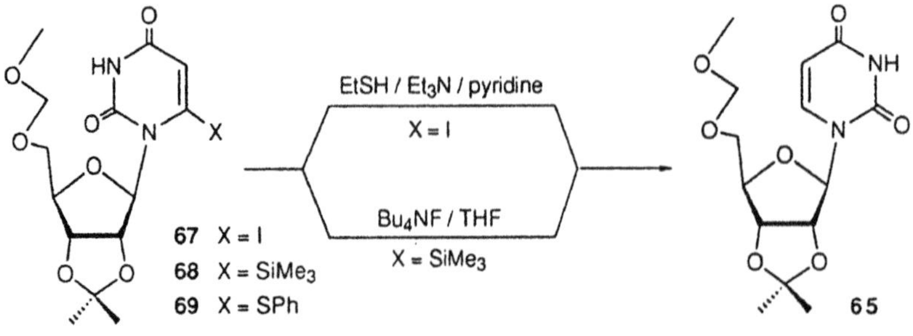

Scheme 13. Conditions for the removal of the 6-iodo and 6-trimethyl-silyl groups.

Lithiation of the 6-phenylthio derivative **69** with butyllithium gave a complex mixture of products, from which a small amount of 6-butyl derivative was isolated as a result of nucleophilic attack of the lithiating agent. Although the use of LDA was ineffective, the use of the more basic LTMP was found to effect almost quantitative lithiation at the 5-position of **69**. As shown in Scheme 14, successive treatment of the lithiated species with electrophiles yielded the 5-substituted 6-(phenylthio)uridines **70–74**.[60] Attempted removal of the 6-phenylthio group in these products with Al-Hg or Raney-Ni under various conditions all met with failure, resulting in either recovery of the starting material or preferential saturation of the 5,6-double bond. However, we found that radical reaction of tributyltin hydride in refluxing benzene can be used to remove the 6-phenylthio group, although its reaction pattern varies from substrate to substrate (Scheme 14).[61] Compounds **70** and **71** underwent desulfurizative stannylation to give the corresponding 5-alkyl-6-tributylstannyl derivatives. In contrast, direct replacement of 6-phenylthio group with hydrogen occurred in the case of the 5-ethoxycarbonyl derivative **72**. Compound **73** remained unchanged, presumably due to the presence of the bulky 5-trimethylsilyl group. In the reaction of compound **74** having two phenylthio groups, the stannylation occurred either at the 5- or 6-position, forming two regioisomers. The desired 5-substituted uridines (**75**) were prepared simply by treating these products with aqueous trifluoroacetic acid.

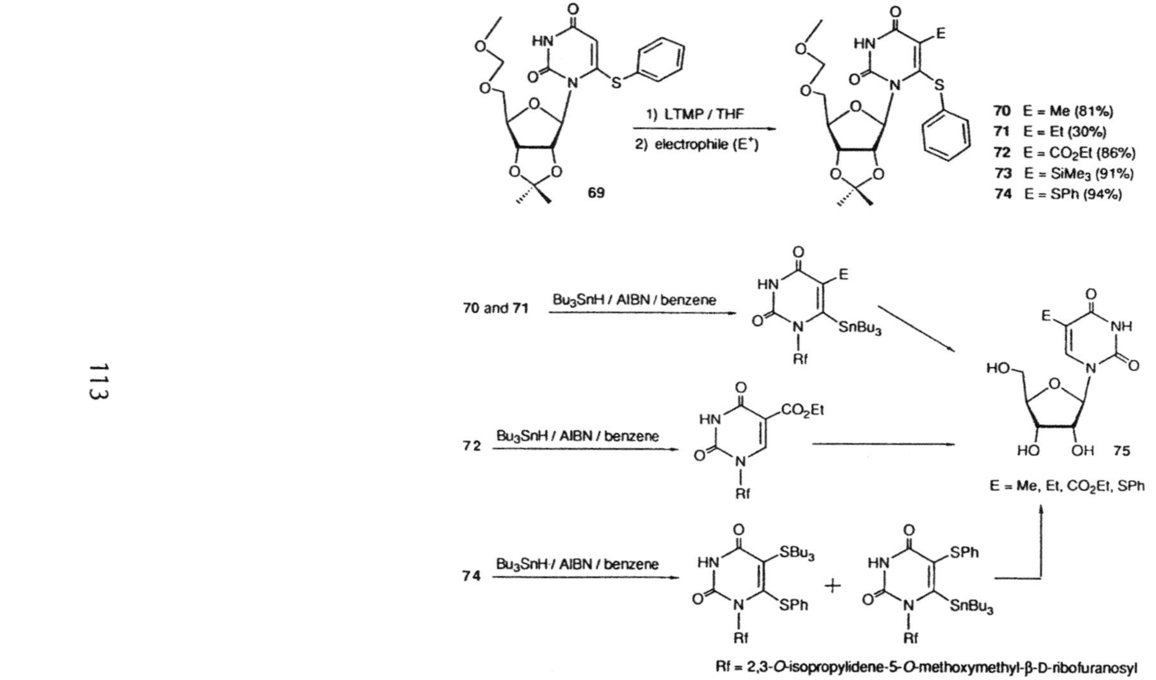

Scheme 14. Synthesis of 5-substituted 6-(phenylthio)uridines (**70–74**) and their transformation to 5-substituted uridines (**75**).

113

Table 1. Chemical Shifts (δ, ppm) of the Two Isopropylidene Methyl Groups of Compounds **65**, **69**, and **70–74** in $CDCl_3$ (100 MHz)

Compd.	5-Substituent	6-Substituent	Exo	Endo
65	H	H	1.36	1.59
69	H	SPh	1.35	1.58
70	Me	SPh	1.23	1.46
71	Et	SPh	1.18	1.40
72	CO_2Et	SPh	1.19	1.45
73	$SiMe_3$	SPh	1.10	1.34
74	SPh	SPh	1.18	1.43

During this study, ^1H NMR evidence was obtained regarding a conformational change of the 6-phenylthio group, which is caused by the presence of the 5-substituent. Although the chemical shifts of the two isopropylidene methyl signals of 2′,3′-O-isopropylidene-5′-O-methoxymethyluridine (**65**) and its 6-phenylthio derivative (**69**) are virtually the same (Table 1), a significant difference was observed in the H-5 resonance between the two compounds (chemical shifts of H-5: **65**, 5.70 ppm vs. **69**, 4.93 ppm), which is certainly due to the magnetic anisotropy of the 6-phenylthio group in the latter case. This indicates that the 6-phenylthio group in **69** lies perpendicular to the uracil ring. When the 5-position is substituted, both *exo* and *endo* methyl signals of the isopropylidene group were shielded by ca. 0.1–0.2 ppm. This observation is clearly attributable to the conformational change of the 6-phenylthio group. There can be seen a trend that the bulkier groups cause the greater change.

C. Effect of the Sugar Protecting Group on Lithiation

LDA lithiation of 5′-O-protected 2′,3′-O-isopropylideneuridine (**65**), as mentioned above, occurs predominantly at the 6-position to the extent of 87–88%. In this case, the amount of LDA used was 2.5 equiv relative to the substrate. The regiospecificity was retained in the lithiation of 2′-deoxy-3′,5′-O-(tetraisopropyldisiloxan-1,3-diyl)uridine (**76**). However, the lithiation level was found to be only 20% when the same amount of LDA was used. Optimization of the reaction conditions led to the highest level of 53% with the use of 5 equiv of LDA. Although some 6-substituted 2′-deoxyuridines (**77**) were syn-

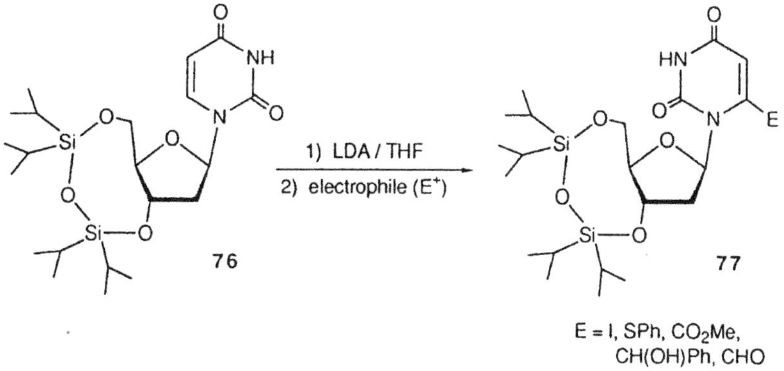

E = I, SPh, CO₂Me,
CH(OH)Ph, CHO

Scheme 15. Synthesis of 6-substituted 2'-deoxyuridine derivatives **77**.

thesized under the optimized conditions,[62] as shown in Scheme 15, there remained a question why such a discrepancy was observed between the lithiation levels of **65** and **76**. This prompted us to reinvestigate the efficiency (and regioselectivity) of the lithiation of uridine by changing the sugar hydroxyl protecting groups.

As shown in Figure 4, compounds **78–81** were selected as substrates. The efficiency and regioselectivity were analyzed by deuterium incorporation.[63] In contradistinction to the result previously mentioned for compound **65**, neither the 5- nor the 6-position of **79** and **80** was lithiated with LDA (lithiation of **78** occurred at the 6-position only to any appreciable extent). In contrast to this, LDA lithiation of the 4-methoxypyrimidin-2-one derivative **81**, the sugar portion of which has the same structure as that of **80**, occurred at the 6-position to the extent of 71%. These results indicate the following: (1) the 2',3'-*O*-isopropylidene protecting group is indispensable for efficient LDA lithiation at the 6-position, (2) the steric environment around the 6-position of **78–80** is more hindered than that in compound **65**, and (3) LDA is not basic enough to deprotonate H-5. A possible explanation for the failure to lithiate the 6-position of **80**, for example, is as follows: the initially formed N3-lithiated species favors a *syn*-conformation which is stabilized by intramolecular chelation as depicted in Scheme 16. In this situation, the approach of LDA would be hindered by the 2'-*O*-TBDMS group.

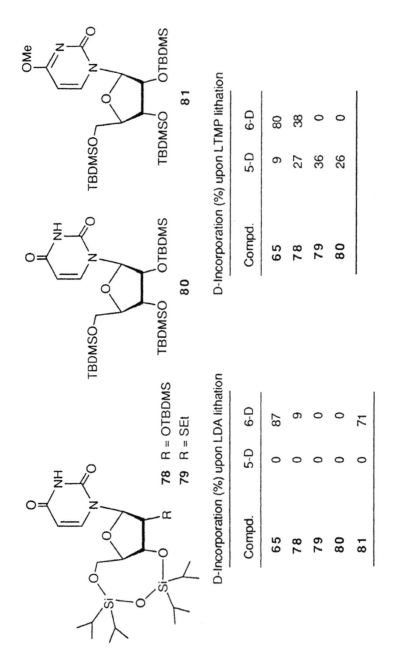

D-Incorporation (%) upon LTMP lithiation

Compd.	5-D	6-D
65	9	80
78	27	38
79	36	0
80	26	0

D-Incorporation (%) upon LDA lithiation

Compd.	5-D	6-D
65	0	87
78	0	9
79	0	0
80	0	0
81	0	71

78 R = OTBDMS
79 R = SEt

Figure 4. The results of deuterium incorporation to compounds **78–81**.

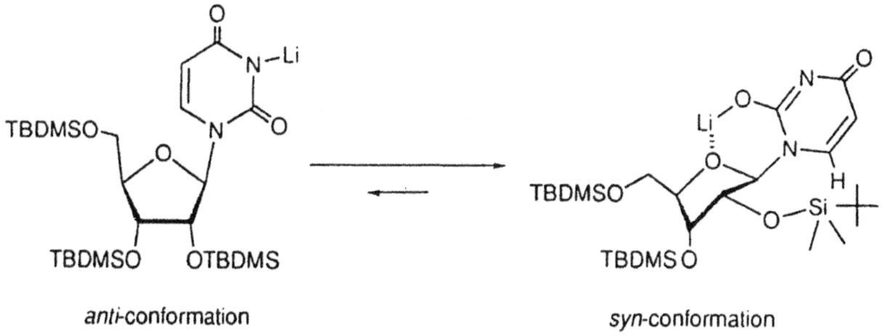

anti-conformation   syn-conformation

Scheme 16. A possible conformation of the N3-lithiated intermediate of compound **81**.

Based on this assumption, selective deprotonation of the less acidic H-5[64] was accomplished in the cases of **79** and **80** by the use of the more basic LTMP (Figure 4). We have no clear explanation about the decreased selectivity in the LTMP lithiation of **78**, but it is certainly consistent with the result observed from its LDA lithiation. Later, a practical lithiation level at the 5-position of **81** with a high regioselectivity (C5, 87% vs. C6, 5%) was attained by a combination of *sec*-butyllithium and *N,N,N′,N′*-tetramethylethylelediamine, and this approach was used for direct entry to 5-substituted uridines.[65,66]

V. HEPT AND ITS ANALOGS

A. Lithiation of Acyclouridines: Discovery of HEPT

The successful clinical use of acyclovir, 9-[(2-hy-droxyethoxy)methyl]guanine, as an excellent antiviral agent has brought about extensive syntheses of a wide variety of acyclic nucleosides modified in either the base moiety or the acyclic structure.[67,68] However, base-modified pyrimidine acyclonucleosides had always been substituted at the 5-position, presumably due to the difficulty of substitution at the 6-position. These facts led us to synthesize 6-substituted acyclouridines as an extension of our lithiation studies of nucleosides. Design of the acyclic structure was derived from acyclovir, and that of the 6-substituent was based on the aforementioned finding that 6-phenylthio and 6-iodouridine derivatives showed antileukemic activity.[59]

Scheme 17 shows the synthetic route to the designed compounds (**84**). One interesting point observed during the LDA lithiation of compounds **82** was that, even in the presence of an electron-donating 5-methyl group, the 6-substituted products **83** were obtained in good yields (isolated yields, 73–79%). It may be concluded, therefore, that the presence of the sugar structure itself exerts a certain degree of steric hinderance in the aforementioned 6-lithiation of 2′,3′-*O*-iso-propylidene-5′-*O*-methoxymethyluridine (**65**).

Compounds **84** were evaluated for their inhibitory activity against HIV-1 (human immunodeficiency virus type 1) replication in MT-4 cells. Among these 10 compounds, 1-[(2-hydroxyethoxy)methyl]-6-(phenylthio)thymine (HEPT: **84**, R = Me, E = SPh) was found to show an inhibitory effect on the cytopathogenecity of HIV-1 (EC_{50} = 7.0 mM).[69] The uracil derivative of HEPT (**84**, R = H, E = SPh) was devoid of such an activity (EC_{50} = >500 mM). The conformational difference of the 6-phenylthio group of these two compounds was analyzed by

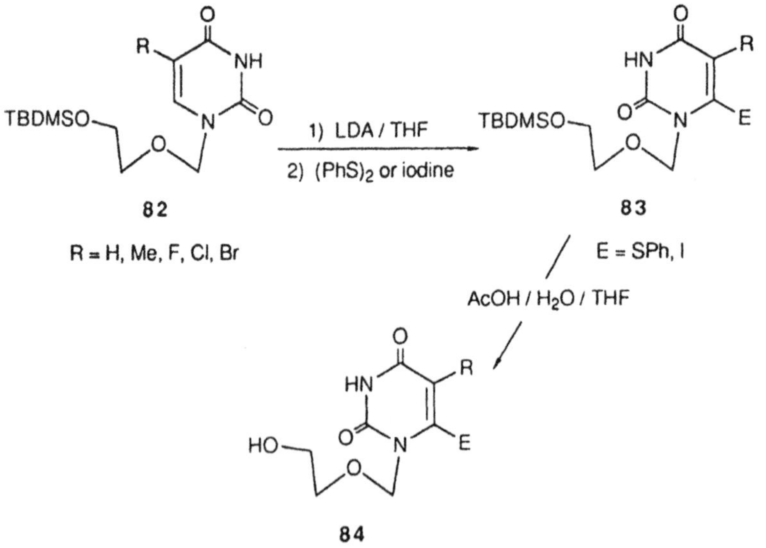

Scheme 17. Synthesis of 6-phenylthio and 6-iodo acyclouridines (**84**).

X-ray crystallography. The result shown in Figure 5 is consistent with our previous ^{1}H NMR findings (Table 1) for the 6-(phenylthio)uridine derivatives.

The anti-HIV-1 activity of HEPT is only marginal when compared with that of AZT and ddC, but HEPT was found to be much less toxic than these nucleoside derivatives (Figure 6). In terms of activity (EC_{50}) and cytotoxicity (CC_{50}), HEPT is almost comparable to ddA. A salient feature of this compound lies in its highly specific antiviral property. When the activity of HEPT was examined by changing the virus as well as the cells, HEPT was uniformly active to various strains of HIV-1, but not to other retroviruses, including HIV-2, and other DNA viruses (Table 2).[70]

Acyclonucleosides can be regarded as mimics of naturally occurring nucleosides, and it was therefore logical to expect antiviral compounds in this class to exhibit activity after being converted to

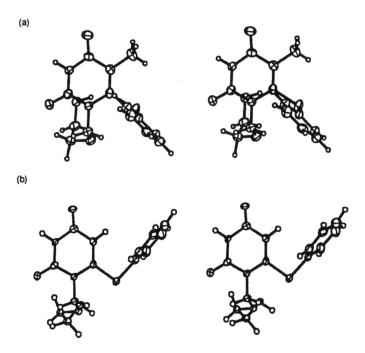

Figure 5. ORTEP stereoviews of (**a**) HEPT and (**b**) its uracil derivative.

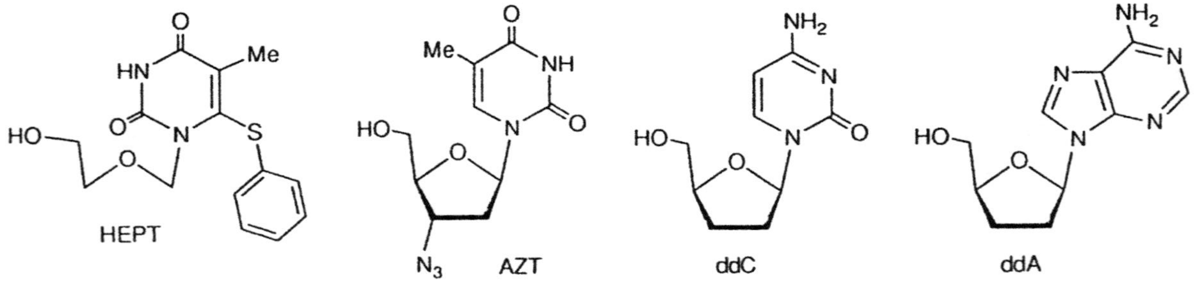

Compd.	EC$_{50}$ (μM)	CC$_{50}$ (μM)	SI
HEPT	7.0	740	106
AZT	0.016	20	1250
ddC	0.3	40	133
ddA	6.3	890	141

EC$_{50}$: effective concentration required to achieve 50% protection of MT-4 cells against the cytopathic effect of HIV-1.
CC$_{50}$: cytotoxic concentration required to reduce viability of mock-infected MT-4 cells by 50%.
SI: selectivity index (ratio of CC$_{50}$/EC$_{50}$).

Figure 6. Anti-HIV-1 (HTLV-III$_B$ strain) activity of HEPT and nucleoside analogs in MT-4 cells.

Table 2. Inhibitory Effect of HEPT on Replication of HIV-1 and
Other Viruses in Cell Cultures

Virus[a]	Cell	EC_{50} (μM)	CC_{50} (μM)
HIV-1 (HTLV-III_B)	MT-4	7.0	740
	CEM (day 4)	1.3	720
	CEM (day 8)	8.3	400
	HUT-78	5.2	>250
	PBL (day 10)	1.6	>250
	PBL (day 16)	2.2	>250
HIV-1 (HTLV-III_{Ba-L})	MP (day 12)	1.0	>100
	MP (day 21)	8.8	>100
HIV-1 (HTLV-III_{RF})	MT-4	12	—
HIV-1 (HE)	MT-4	16	—
HIV-2 (ROD)	MT-4	>250	—
	CEM (day 4)	>250	—
HIV-2 (EHO)	MT-4	>250	—
SIV_{MAC}	MT-4	>500	—
SRV	Raji	>500	>500
MSV	C3H	>300	>300
HSV-1 (KOS)	PRK	>300	>300
HSV-2 (G)	PRK	>300	—
Vaccinia virus	PRK	>300	—
VSV	HeLa	>600	>600
Coxsackie virus type B4	HeLa	>600	—
Poliovirus type 1	HeLa	>600	—
Parainfluenza virus type 3	Vero	>300	>300
Reovirus type 1	Vero	>300	—
Sindbis virus	Vero	>300	—
Semliki forest virus	Vero	>300	—

Note: [a]Abbreviations: HIV, human immunodeficiency virus; SIV, simian
immunodeficiency virus; SRV, simian related virus; MSV, murine (moloney)
sarcoma virus; HSV, herpes simplex virus; VSV, vesicular stomatitis virus.

their monophosphates and then to triphosphates, as reported in the
case of the anti-HSV agent acyclovir.[71] However, HEPT has the
following two unique properties: (1) HEPT does not compete with
thymidine for phosphorylation by thymidine kinase derived from
MT-4 cells, and (2) the synthetic triphosphate of HEPT does not inhibit
HIV-1 reverse transcriptase (RT) at concentrations much higher than
that of its EC_{50}, irrespective of the template primer used [either
poly(A):oligo(dT) or poly(C):oligo(dG)]. These observations suggest

that, although HEPT is structurally an acyclonucleoside, this compound may be acting as an anti-HIV-1 agent without being phosphorylated. Later, emergence of structurally diverse non-nucleoside reverse transcriptase inhibitors (NNRTIs)[72] (such as TIBO,[73] nevirapine,[74] L-696229,[75] TSAO,[76] loviride,[77] PETT,[78] L-743726,[79] TDA[80] etc.) revealed that HEPT can be classified as a NNRTI.

B. Structure–Activity Relationships of HEPT Analogs

Having been encouraged by the finding of an anti-HIV-1 specific lead HEPT, synthesis of its analogs was carried out with the aim to improve its activity.[81–92] The synthetic methods used here are: (1) LDA lithiation for the preparation of various types of 6-modified analogs,[93] (2) LTMP lithiation for the modification at the 5-position of a 6-phenylthio derivative, and (3) an addition–elimination reaction[81] of a 6-phenylsulfinyl derivative. The third method provides an entry to analogs having oxygen- or nitrogen-containing substituents at the 6-position,[94,95] which are difficult to prepare by the lithiation strategy.

Modification at the 6-position of HEPT such as the introduction of a simple alkylthio (SMe, SBu, SBu-*t*) group resulted in a complete loss of activity. Although replacement of the 6-phenylthio group with a phenoxy or phenylamino group turned out to be ineffective, it is interesting to see that the 6-cyclohexylthio (**85**) and the 6-benzyl (**86**) analogs retained activity (Figure 7). Since compounds **87** and **88** were inactive, the presence of a ring structure at the 6-position separated from the heterocycle by one atom seems to be necessary for anti-HIV-1 activity. From the activity of compounds variously modified (I, SPh, alkynyl, alkenyl, acyl, benzyl, allyl, alkyl, etc.) at the 5-position of HEPT, it became apparent that a bulkier 5-substituent potentiates the anti-HIV-1 activity as can be seen from the data of compounds **89** and **91** in Figure 7. This was thought to be due to the greater conformational change of the 6-phenylthio group.

The effect of substitution in the 6-phenylthio ring was also examined, and the results are summarized in Table 3. Although *ortho* substitution is slightly better than *para* substitution, both resulted in a significant decrease in the original activity of HEPT. On the other hand, most *meta* substitutions, except those with a highly polar functional group, appeared to be beneficial modifications. One noteworthy

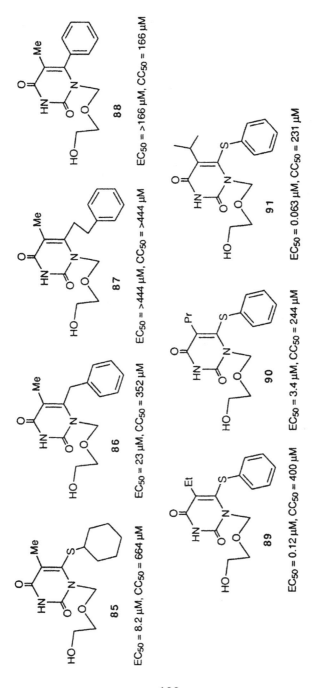

85 EC$_{50}$ = 8.2 μM, CC$_{50}$ = 664 μM

86 EC$_{50}$ = 23 μM, CC$_{50}$ = 352 μM

87 EC$_{50}$ = >444 μM, CC$_{50}$ = >444 μM

88 EC$_{50}$ = >166 μM, CC$_{50}$ = 166 μM

89 EC$_{50}$ = 0.12 μM, CC$_{50}$ = 400 μM

90 EC$_{50}$ = 3.4 μM, CC$_{50}$ = 244 μM

91 EC$_{50}$ = 0.063 μM, CC$_{50}$ = 231 μM

Figure 7. The anti-HIV-1 activity of HEPT analogs modified either at the 6- or the 5-position.

Table 3. Effect of Modification of the 6-Phenylthio Ring of HEPT on Anti-HIV-1 Activity

Position	Substituent	EC_{50} (μM)	CC_{50} (μM)	SI
ortho	Me	71	>250	>3.5
	Cl	>130	130	<1
	NO_2	140	>250	>1.8
	OMe	19	>250	>13
para	Me	220	>250	>1.1
	F	>250	>250	
	Cl	>250	>250	
	NO_2	>190	190	<1
	OMe	>250	>250	
meta	Me	2.6	420	162
	Et	2.7	181	67
	Bu-t	12	75	6.3
	CH_2OH	>292	292	<1
	CF_3	45	196	4.4
	F	3.3	282	85
	Cl	13	210	16
	Br	5.7	141	25
	I	10	106	11
	NO_2	34	170	5.0
	OH	82	446	5.3
	OMe	22	>250	>11
	CO_2Me	7.9	221	28
	COME	7.3	228	35
	CO_2H	>352	352	<1
	$CONH_2$	>306	306	<1
	CN	10	234	23
	di-Me	0.26	243	935
	di-Cl	1.3	130	110
HEPT		7.0	743	106

result can be seen in the double *meta* substitution with methyl groups or with chlorine atoms. Especially in the former case, a substantial increase in the activity led to an SI value which is 9-fold greater than that of HEPT.

Compounds **92–96** depicted in Figure 8 were synthesized to investigate the effect of other modifications in the base moiety of HEPT. Their EC_{50} values indicate that the only effective modification is the

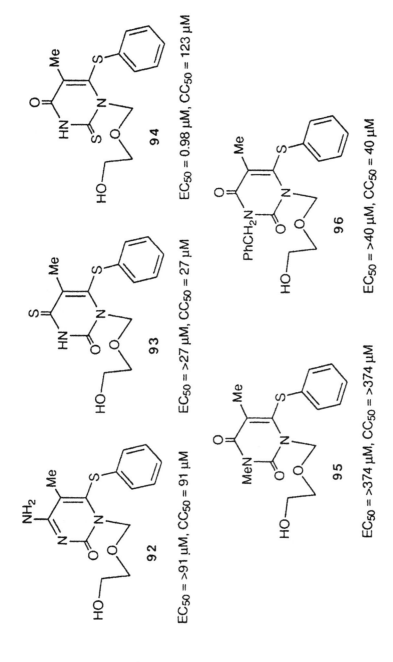

94

EC$_{50}$ = 0.98 μM, CC$_{50}$ = 123 μM

93

EC$_{50}$ = >27 μM, CC$_{50}$ = 27 μM

92

EC$_{50}$ = >91 μM, CC$_{50}$ = 91 μM

96

EC$_{50}$ = >40 μM, CC$_{50}$ = 40 μM

95

EC$_{50}$ = >374 μM, CC$_{50}$ = >374 μM

Figure 8. The anti-HIV-1 activity of HEPT analogues variously modified in the base moiety.

transformation to the 2-thio derivative (**94**). The lack of activity of the N3-alkylated analogs (**95** and **96**) suggests that the presence of N3-H is essential for designing active analogs of HEPT.

As mentioned earlier, experimental results suggest that there would be no phosphorylation step involved during the antiviral action of HEPT against HIV-1. One would readily assume, therefore, that a hydroxyl group in the acyclic structure of HEPT would not contribute to its inhibitory activity against HIV-1 replication. As shown in Figure 9, compound **97** prepared by a simple acetylation of HEPT gave an almost comparable EC_{50} value to that of HEPT. This is also the case for the more biologically stable methyl ether **98**. These preliminary results led to the synthesis of a series of "5'-deoxy" analogs, such as compounds **99–103**. The genuine deoxy analog **99** was found to be 20-fold more active than the lead compound HEPT. There can be seen a trend that the value of the EC_{50} correlates with the size of the acyclic side chain, with the exception of the benzyloxymethyl derivative **102**. It should be mentioned that compound **103**, which bears the simplest alkoxymethyl group, has a higher activity than HEPT itself. The anti-HIV-1 activity of HEPT analogs can be retained even upon removal of the ether oxygen in the acyclic structure, as evidenced by the EC_{50} values of the N1-alkyl analogs **104** and **105**.[96] It seems likely that introduction of an alkyl group longer than ethyl group is necessary, since 6-(phenylthio)thymine itself and its N1-methyl derivative did not show any activity.

Through the above extensive synthetic studies, some structure–activity relationships of HEPT analogs have become apparent.[97] The factors which enhance or retain the original activity of HEPT are (1) substitution of the 6-phenylthio ring at the *meta* position with one or two lipophilic groups such as a methyl group, (2) replacement of the 6-phenylthio substituent with a benzyl group, (3) replacement of the 2-oxo function by a 2-thione function, (4) replacement of the methyl group at the 5-position with a bulkier group such as the ethyl or isopropyl group, and (5) modification of the acyclic structure.

C. MKC-442: A Clinical Candidate for AIDS Chemotherapy

A series of highly active HEPT analogs were designed and synthesized based on the available structure–activity relationships. In Figure

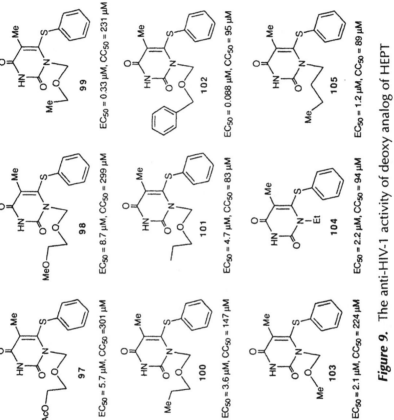

99 EC$_{50}$ = 0.33 μM, CC$_{50}$ = 231 μM

98 EC$_{50}$ = 8.7 μM, CC$_{50}$ = 299 μM

97 EC$_{50}$ = 5.7 μM, CC$_{50}$ = 301 μM

102 EC$_{50}$ = 0.088 μM, CC$_{50}$ = 95 μM

101 EC$_{50}$ = 4.7 μM, CC$_{50}$ = 83 μM

100 EC$_{50}$ = 3.6 μM, CC$_{50}$ = 147 μM

105 EC$_{50}$ = 1.2 μM, CC$_{50}$ = 89 μM

104 EC$_{50}$ = 2.2 μM, CC$_{50}$ = 94 μM

103 EC$_{50}$ = 2.1 μM, CC$_{50}$ = 224 μM

Figure 9. The anti-HIV-1 activity of deoxy analog of HEPT

10 the inhibitory effect of some of these active compounds on HIV-1 replication is summarized. Compounds **106–113** have now a comparable EC_{50} value to that of AZT. The observed EC_{50} values also indicate that each factor, deduced from the structure–activity relationship studies, is at work cooperatively in improving anti-HIV-1 activity. It is therefore reasonable to assume that HEPT analogs exert their activity without being subjected to any metabolic activation. Reverse transcriptase (RT) assays have revealed that HEPT analogs act specifically on HIV-1 RT.[91,92] Compound **109**, for example, was found to be inhibitory againt HIV-1 RT, irrespective of the source of the enzyme and template primer used (Table 4). However, this compound did not affect the activity of HIV-2 at concentrations up to 500 mM.

The data given in Figure 10 were obtained by performing the anti-HIV-1 assay in the presence of 10% heat-inactivated fetal bovine serum (FBS). When the assays were performed under more therapeutically oriented conditions by adding human serum (HS), some compounds were found to be less inhibitory to HIV-1 replication due to their binding to human serum albumin.[98] This decrease in activity correlates with their lipophilicity, and is particularly notable in compounds having the 6-phenylthio group and/or the benzyloxymethyl acyclic moiety. Also, a preliminary pharmacokinetic study of some HEPT analogs suggested that, although *meta* substitution at the 6-phenylthio (or 6-benzyl) ring potentiates the activity, the resulting compounds would suffer from considerably lower bioavailability.[90] By taking these facts into consideration, MKC-442 (**113**) was selected from these compounds as a candidate for clinical use.[99]

MKC-442 is inhibitory not only to AZT-susceptible strains of HIV-1 but also to AZT-resistant mutants. Its anti-HIV-1 (HTLV-III$_B$) activity evaluated in MT-4 cells is approximately 5-fold higher than that of the recently approved NNRTI nevirapine.[99a] Like other HEPT analogs,[100] the combination of MKC-442 with either AZT, ddC, or ddI synergistically inhibits the replication of HIV-1. Although this is a common property seen for other NNRTIs, MKC-442 is unique in that it shows synergism with AZT 5′-triphosphate at the HIV-1 RT level. No such synergism was observed for TIBO, nevirapine, or L-696229.[101]

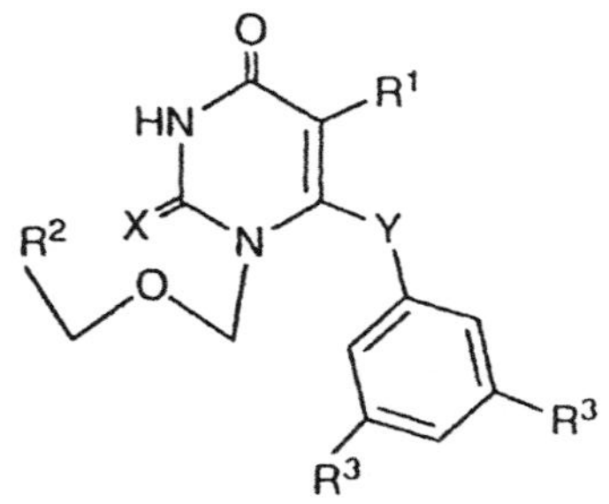

106-113

Compd.	R¹	R²	R³	X	Y	EC$_{50}$ (μM)	CC$_{50}$ (μM)	SI
106	Et	CH$_2$OH	Me	S	S	0.0075	172	23000
107	Et	Me	Me	O	S	0.0062	>100	>16100
108	Et	Ph	H	O	S	0.0049	30	6100
109	Et	Me	Me	O	CH$_2$	0.0022	249	113000
110	i-Pr	CH$_2$OH	Me	O	S	0.0027	128	47400
111	i-Pr	Ph	Me	S	S	0.0068	>20	>2940
112	i-Pr	Me	Me	O	CH$_2$	0.0006	43	72000
113	i-Pr	Me	H	O	CH$_2$	0.0042	186	44000
HEPT	Me	CH$_2$OH	H	O	S	7.0	740	106
AZT						0.003	7.8	2600

Figure 10. The inhibitory effect of highly active HEPT analogues on HIV-1 (HTLV-III$_B$) replication in MT-4 cells.

Table 4. Inhibitory Effects of Compound **109** on HIV RT Activity

Enzyme	Template-Primer	Substrate	IC_{50} (μM)
HIV-1 recombinant RT-A[a]	poly(A):oligo(dT)	TTP	0.11
	poly(C):oligo(dG)	dGTP	0.044
HIV-1 recombinant RT-B[b]	poly(A):oligo(dT)	TTP	0.16
	poly(C):oligo(dG)	dGTP	0.036
HIV-1 native RT	poly(A):oligo(dT)	TTP	0.36
HIV-2 native RT	poly(A):oligo(dT)	TTP	>500
	poly(C):oligo(dG)	dGTP	>500

Notes: [a]Obtained from National Institute of Allergy and Infectious Diseases.
[b]Obtained from MicroGeneSys.

D. Molecular Mechanism of Interactions between HIV-1-RT and HEPT Analogs

The efficacy of NNRTIs can only be improved rationally when detailed structural data on the protein and its interaction with such inhibitors can be analyzed in detail. Thus, following the first publication of a crystal structure of HIV-1 RT complexed with nevirapine,[102] several publications have addressed this problem.[103–107]

Data on compounds from three chemically unrelated groups of NNRTIs showed that these inhibitors bound at equivalent sites with a high degree of spatial overlap. Given the chemical diversity of the inhibitors, the shapes of the bound inhibitors were surprisingly similar. Although it can be argued that crystal packing forces may be a significant factor in the structures seen in such a flexible enzyme, studies have shown,[107] that loss and rebinding of NNRTIs can occur without significant domain rearrangement and indicate a common mechanism for inhibition through NNRTI-induced distortion of the polymerase active site. Independent kinetic studies[110–112] had previously suggested a similar effect which resulted in the required Mg^{+2} ions not being in proper alignment with the carboxyl groups of the conserved aspartic acid residues in the presence of an NNRTI. This interfered with catalysis at the polymerization site; thus, catalysis, rather than binding, seems to be affected.

Although informative, the structural experiments carried out with a diversity of NNRTIs only allowed general conclusions to be drawn,

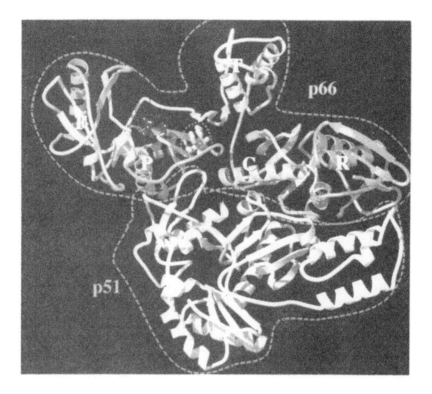

Figure 11. The HIV-1 RT heterodimer, as seen complexed with MKC-442. The orientation is chosen such that the p66 subunit, which carries the single polymerase active site is at the top of the figure. The domains are labeled for the p66 subunit: F-fingers, P-palm, T-thumb, C-connection, and R-RNase H. Three conserved catalytic aspartic acid residues define the polymerase catalytic site, whereas MKC-442 defines the position of the NNRTI binding pocket. Drawn with Bobscript (R. Esnouf, unpublished), a development of Molscript[114] and rendered with Raster3D.[115]

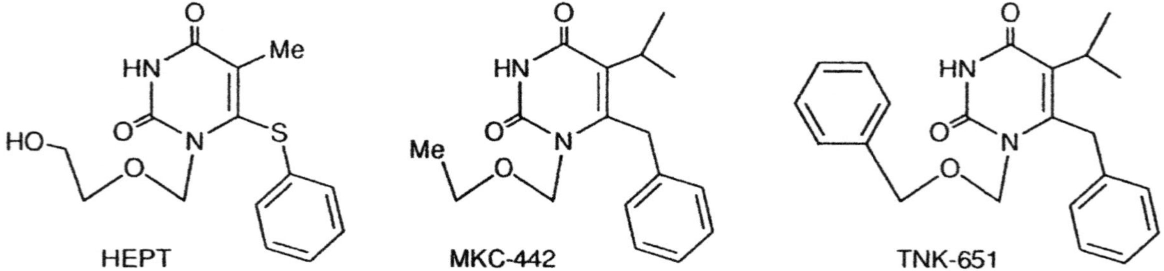

Figure 12. Structures of HEPT, MKC-442, and TNK-651.

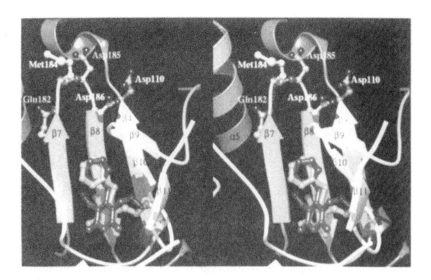

Figure 13. Stereographic illustration of the NNRTI binding pocket showing its relationship to the polymerase active site. Key conserved amino acid side chains in the polymerase active site are shown in all-atom representation (and labeled), elsewhere the conventional spiral (for α-helices) and arrow (for β-sheet) representation is used (elements of secondary structure within the protein are labeled using the convention of Ren et al.[104] HEPT and MKC-442 are in the NNRTI binding pocket. The structures were separately superposed onto the structure of the RT–nevirapine complex to compensate for differences in crystal form and domain orientations, in the manner described previously.[113] Drawn as described in the legend to Figure 11.

and a recent manuscript[113] has focussed on a comparison of the differences between compounds from the HEPT series (Figures 11 and 12): HEPT itself and MKC-442 (or TNK-651) which differ in efficacy by a factor of 10^3 despite having very similar chemical structures.

The spectrum of drug resistance mutations seen for two of the analogs is also different; HEPT resistance is acquired by the presence of a single Tyr188His mutation,[114] whereas MKC-442 resistance requires Tyr181Cys and either Lys103Arg or Val108Ile mutations (Figures 13–16).[115]

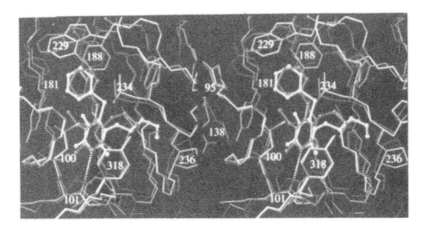

Figure 14. Stereo diagram showing the structure of the HEPT-RT and MKC-442-RT complexes in the region of the bound inhibitor. The conserved hydrogen bond is between the N3-H of the inhibitors and the carbonyl oxygen of residue **101** of the enzyme. The superpositions were performed as described in the legend to Figure 13. Drawn as described in the legend to Figure 11.

The structures of the protein-inhibitor complexes have been obtained at a resolution of 2.55 Å and the results confirm that the HEPT analogs bind to an allosteric site some 10 Å from the polymerase catalytic site. Comparison of the structures makes it possible to understand why small changes in chemical structure, particularly at the 5-position of the pyrimidine ring, lead to a significant alteration in potency. The two key features seen are a discrete conformational switch of Tyr181 and a continuous variation in the position of the Pro236 hairpin. The most obvious structural difference between the complex with HEPT and with MKC-442 is the conformation of the side chain of Tyr181 which in the latter case has swung such that the aromatic ring now interacts with the 6-benzyl ring of the inhibitor and other nearby aromatic residues in the protein. The binding of HEPT only significantly affects Tyr181 and Tyr188 and it is likely that this difference explains the considerable difference in binding affinities

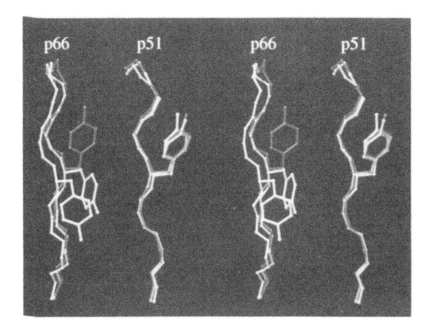

Figure 15. Comparison of the conformation of residue Tyr181 in unliganded, HEPT bound, and MKC-442 bound structures. The conformations in the p66 and p51 chains are shown alongside each other (p66 on the left) in this stereographic representation. The superpositions were performed as described in the legend to Figure 13 with the addition of each chain and some disordered loops (residues 97–118, 149–209, and 234–240). Drawn as described in the legend to Figure 11.

between MKC-442 and HEPT. The change in orientation of Tyr181 is triggered by the steric interactions between the larger 5-isopropyl group in MKC-442 and the neighboring protein which force Tyr181 into a conformation found in the p51 subunit, which does not show any catalytic activity.

HIV-RT is a flexible protein and can accommodate a variety of structures based on HEPT, many of which are highly efficient inhibitors. Differences in the acyclic moiety between MKC-442 and TNK-651 (ethoxymethyl vs. bezyloxymethyl) can be accommodated by the flexibility of the Pro236 hairpin. Binding of HEPT-NNRTIs to HIV-

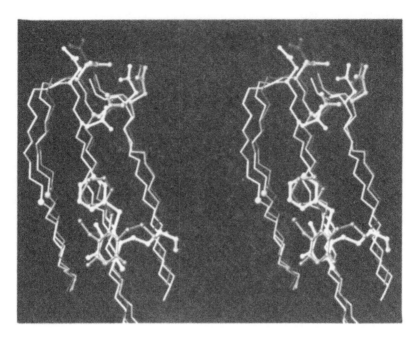

Figure 16. Comparison of the β-strands bearing the catalytic aspartic acid residues of the p66 subunit of HIV-1 RT between the apo enzyme and the inhibited enzyme containing HEPT and MKC-442. The super-position was performed on the basis of the NNRTI-binding sites (as for Figure 13). Note that the repositioning of the active site residues is essentially the same for both inhibitor bound complexes. For orientation purposes the positions of the Cα atoms of residue Tyr181 are shown as spheres on the three backbone traces. Drawn as described in the legend to Figure 11.

RT causes a repositioning of the catalytic residues Asp110, 185, and 186. The results suggest that high potency compounds result from a tight binding which causes the conformational change in Tyr181. In all the complexes however, the displacement of the catalytic aspartic acid residues is similar; it just requires a 1000-fold increase in concentration of HEPT compared to MKC-442 to achieve the same result.

These structural studies have resulted in a rational explanation of how a potent inhibitor is more effective than a closely related but less potent analog. They also explain why HEPT analogs are effective only against HIV-1 and not against HIV-2 which lacks both Tyr181 and Tyr188. They also explain how resistance to MKC-442 is achieved primarily by the Tyr181Cys mutation and they suggest that there is considerable potential to exploit inhibitor-protein interactions in the design of ever more potent inhibitors.

VI. CONCLUSION

The research covered by this review, which started with a fundamental investigation into the lithiation of pyrimidine nucleosides and hence the ability to derivatize the 5- and 6-positions of uracil analogs, led to the serendipitous discovery of a lead compound with activity against HIV-1. Structure-activity relationship studies have improved the efficacy by a factor of 10^4 and the compounds were shown to interact directly with HIV-1 reverse transcriptase. X-ray structural studies of inhibitor-enzyme complexes have shown how the inhibitors work and have suggested new derivatives to be synthesized which might have increased potency.

ACKNOWLEDGMENTS

The synthesis of a vast number of HEPT analogs has been accomplished by a team of skilled chemists (Dr. Masaru Ubasawa, Mr. Hideaki Takashima, and Mr. Kouichi Sekiya) at Mitsubishi Chemical Corporation Research Center. Part of this work has been supported by a joint research project between JSPS and Royal Society-British Council (to H. T., D. I. S., and R. T. W.). The authors (H. T. and R. T. W.) acknowledge Drs. John Richards, John Grote, and Robin Sowden (the science officers, Tokyo) of the British Council for their continuing interest and support. One of the authors (H. T.) thanks also Miss Miyoko Hayashi for her excellent secretarial help.

REFERENCES AND NOTES

1. Jones, R. G.; Gilman, H. In *Organic Reactions*; Adams, R.; Adkins, H.; Blatt, A. H.; Cope, A. C.; McGrew, F. C.; Niemann, C.; Snyder, H. R., Eds.; John Wiley and Sons: New York, 1951,Vol. 6, p 339.
2. Parham, W. E.; Bradsher, C. K. *Acc. Chem. Res.* **1982**, *15*, 300.

3. Gilman, H.; Jacoby, A. L. *J. Org. Chem.* **1938**, *3*, 108.
4. Gilman, H.; Bebb, R. L. *J. Am. Chem. Soc.* **1939**, *61*, 109.
5. Wittig, G.; Fuhrmann, G. *Chem. Ber.* **1940**, *73*, 1197.
6. Gilman, H.; Morton, J. W., Jr. In *Organic Reactions*; Blatt, A. H.; Cope, A. C.; Curtin, D. Y.; McGrew, F. C.; Niemann, C., Eds.; John Wiley and Sons; New York, 1954, Vol. 8, p 258.
7. Beak, P.; Snieckus, V. *Acc. Chem. Res.* **1982**, *15*, 306.
8. Gschwend, H. W.; Rodriguez, H. R. In *Organic Reactions*; Dauben, W. G.; Boswell, G. A., Jr.; Danishefski, S.; Heck, R. F.; Hirschmann, R. F.; Kende, A. S.; Paquette, L. A.; Posner, G. H.; Trost, B. M., Eds.; John Wiley and Sons: New York, 1979, Vol. 26, p 1.
9. Ueda, T. In *Chemistry of Nucleosides and Nucleotides*; Townsend, L., Ed.; Plenum Press: New York, 1988, Vol. 1, p 1.
10. Srivastava, P. C.; Robins, R. K.; Meyer, R. B., Jr. In *Chemistry of Nucleosides and Nucleotides;* Townsend, L., Ed.; Plenum Press: New York, 1988, Vol. 1, p 113.
11. Ulbricht, T. L. V. *Tetrahedron* **1959**, *6*, 225.
12. Pichat, L.; Godbillon, J.; Herbert, M. *Bull. Chim. Soc. Fr.* **1973**, 2712.
13. There are some inconsistencies between the results in ref. 12 and those reported later. In the following reports, no evidence for the formation of the 6-litho intermediate was obtained and there were isolated 2'-deoxyuridine and 2'-deoxy-5-(trimethylsilyl)-uridine as by-products: (a) Schinazi, R. F.; Prusoff, W. H. *Tetrahedron Lett.* **1978**, 4981. (b) Coe, P. L.; Harnden, M. R.; Jones, A. S.; Noble, S. A.; Walker, R. T. *J. Med. Chem.* **1982**, *25*, 1329. (c) Schinazi, R. F.; Prusoff, W. H. *J. Org. Chem.* **1985**, *50*, 841.
14. The halogen-lithium exchange of trimethylsilylated 8-bromopurine nucleosides has also been reported: Công-Danh, N.; Beaucourt, J.-P.; Pichat, L. *Tetrahedron Lett.* **1979**, 2385.
15. Pichat, L.; Godbillon, J.; Herbert, M. *Bull. Chim. Soc. Fr.* **1973**, 2715.
16. Barton, D. H. R.; Hedgecock, C. J. R.; Lederer, E.; Motherwell, W. B. *Tetrahedron Lett.* **1979**, 279.
17. Tanaka, H.; Uchida, Y.; Shinozaki, M.; Hayakawa, H.; Matsuda, A.; Miyasaka, T. *Chem. Pharm. Bull.* **1983**, *31*, 787.
18. It has been reported that 6-chloro-9-(tetrahydropyran-2-yl)purine undergoes C8-lithiation with butyllithium (in THF at −78 °C) and, upon treatment with acetone, gave 6-chloro-8-(2-hydroxy-2-propyl)-9-(tetrahydropyran-2-yl)purine in 70% yield: Leonard, N. J.; Bryant, J. D. *J. Org. Chem.* **1979**, *44*, 4612.
19. Hayakawa, H.; Haraguchi, K.; Tanaka, H.; Miyasaka, T. *Chem. Pharm. Bull.* **1987**, *35*, 72.
20. For 8-chlorination of naturally occurring purine nucleosides, see: Hayakawa, H.; Tanaka, H.; Haraguchi, K.; Mayumi, M.; Nakajima, M.; Sakamaki, T.; Miyasaka, T. *Nucleosides Nucleotides* **1988**, *7*, 121.

21. Hayakawa, H.; Tanaka, H.; Sasaki, K.; Haraguchi, K.; Saitoh, T.; Takai, F.; Miyasaka, T. *J. Heterocyclic Chem.* **1989**, *26*, 189.
22. Kato, K.; Hayakawa, H.; Tanaka, H.; Kumamoto, H.; Miyasaka, T. *Tetrahedron Lett.* **1995**, *36*, 6507.
23. (a) Tius, M. A.; Kawakami, J. K. *Synth. Commun.* **1992**, *22*, 1461. (b) Tius, M. A. *Tetrahedron* **1995**, *51*, 6605.
24. For a review for Stille reaction, see: Mitchell, T. N. *Synthesis* **1992**, 803.
25. The benzoylation of **24** to form **36** was effected according to the published procedure without using a palladium catalyst: Yamamoto, Y.; Yanagi, A. *Chem. Pharm. Bull.* **1982**, *30*, 2003.
26. Maeda, M.; Nushi, K.; Kawazoe, Y. *Tetrahedron* **1974**, *30*, 2677.
27. Matsuda, A.; Nomoto, Y.; Ueda, T. *Chem. Pharm. Bull.* **1979**, *27*, 183.
28. (a) Matsuda, A.; Shinozaki, M.; Miyasaka, T.; Machida, H.; Abiru, T. *Chem. Pharm. Bull.* **1985**, *33*, 1766. (b) Matsuda, A.; Shinozaki, M.; Yamaguchi, T.; Homma, H.; Nomoto, R.; Miyasaka, T.; Watanabe, Y.; Abiru, T. *J. Med. Chem.* **1992**, *35*, 241.
29. (a) Nair, V.; Lyons, A. G. *Tetrahedron* **1990**, *46*, 7677. (b) Nair, V.; Purdy, D. F. ibid. **1991**, *47*, 365. (c) Adah, S. A.; Nair, V. *Tetrahedron Lett.* **1995**, *36*, 6371.
30. (a) Nair, V.; Richardson, S. G. *Synthesis* **1982**, 670. (b) Nair, V.; Young, D. A. *J. Org. Chem.* **1985**, *50*, 406.
31. For a review concerning nucleoside antibiotics, see: Isono, K. *J. Antibiotics* **1988**, *41*, 1711.
32. (a) Kato, K.; Hayakawa, H.; Tanaka, H.; Kumamoto, H.; Shindoh, S.; Shuto, S.; Miyasaka, T. *J. Org. Chem.* **1997**, *62*, 6833. (b) Kumamoto, H.; Hayakawa, H.; Tanaka, H.; Shindoh, S.; Kato, K.; Miyasaka, T.; Endo, K.; Machida, H.; Matsuda, A. *Nucleosides Nucleotides* **1997**, *16*, 15.
33. (a) Mizuno, K.; Tsujino, M.; Takada, M.; Hayashi, M.; Atsumi, K.; Asano, K.; Matsuda, T. *J. Antibiot.* **1974**, *27*, 775. (b) Fukukawa, K.; Shuto, S.; Hirano, T.; Ueda, T. *Chem. Pharm. Bull.* **1984**, *32*, 1644. (c) Idem, ibid., **1986**, *34*, 3653.
34. (a) Reepmeyer, J. C.; Kirk, K. L.; Cohen, L. A. *Tetrahedron Lett.* **1975**, 4107. (b) Wyss, P. C.; Fischer, U. *Helv. Chim. Acta* **1978**, *61*, 3149. (c) Cook, P. D.; Rousseau, R. J.; Mian, A. M.; Dea, P.; Meyer, R. B., Jr.; Robins, R. K. *J. Am. Chem. Soc.* **1976**, *98*, 1492. (d) Alonso, R.; Andrés, J. I.; García-López, M. T.; de las Heras, F. G.; Herranz, R.; Alarcón, B.; Carrasco, L. *J. Med. Chem.* **1985**, *28*, 834. (e) Wood, S. G.; Upadhya, K. G.; Dalley, N. K.; McKernan, P. A.; Canonico, P. G.; Robins, R. K.; Revankar, G. R. *J. Med. Chem.* **1985**, *28*, 1198.
35. Srivastava, P. C.; Streeter, D. G.; Matthews, T. R.; Allen, L. B.; Sidwell, R. W.; Robins, R. K. *J. Med. Chem.* **1976**, *19*, 1020.
36. Synthesis of the 5-iodo derivative from **46** and its palladium-catalyzed transformation to the 5-alkynyl derivatives have been reported: (a) Matsuda, A.; Minakawa, N.; Sasaki, T.; Ueda, T. *Chem. Pharm. Bull.* **1988**, *36*, 2730. (b)

Minakawa, N.; Takeda, T.; Sasaki, T.; Matsuda, A.; Ueda, T. *J. Med. Chem.* **1991**, *34*, 778.

37. For an example, see: Noyce, D. S.; Stowe, G. T. *J. Org. Chem.* **1973**, *38*, 3762.
38. Otter, B. A.; Falco, E. A.; Fox, J. J. *J. Org. Chem.* **1969**, *34*, 1390 and 2636.
39. Tanaka, H.; Takahashi, T.; Togashi, H.; Ueda, T. *Chem. Pharm. Bull.* **1978**, *26*, 3322.
40. (a) Tanaka, H.; Hirayama, M.; Matsuda, A.; Miyasaka, T.; Ueda, T. *Chem. Lett.* **1985**, 589. (b) Tanaka, H.; Hirayama, M.; Suzuki, M.; Miyasaka, T.; Matsuda, A.; Ueda, T. *Tetrahedron* **1986**, *42*, 1971.
41. Allen, L. B.; Huffman, J. H.; Cook, P. D.; Meyer, R. B., Jr.; Robins, R. K.; Sidwell, R. W. *Antimicrob. Agents Chemother.* **1977**, *12*, 114.
42. Synthesis of 3-deazapurine nucleosides has been achieved by ring closure of 5-ethynyl-1-β-D-ribofuranosylimidazole-4-carboxamide or -carbonitrile: (a) Minakawa, N.; Matsuda, A. *Tetrahedron* **1993**, *49*, 557. (b) Minakawa, N.; Sasabuchi, Y.; Kiyosue, A.; Kojima, N.; Matsuda, A. *Chem. Pharm. Bull.* **1996**, *44*, 288.
43. Suzuki, M.; Tanaka, H.; Miyasaka, T. *Chem. Pharm. Bull.* **1987**, *35*, 4056.
44. Tanaka, H.; Nasu, I.; Miyasaka, T. *Tetrahedron Lett.* **1979**, 4755.
45. For an improved method of the synthesis of 6-methyluridine, see: Tanaka, H.; Hayakawa, H.; Shibata, S.; Haraguchi, K.; Miyasaka, T.; Hirota, K. *Nucleosides Nucleotides* **1992**, *11*, 319.
46. Tanaka, H.; Nasu, I.; Hayakawa, H.; Miyasaka, T. *Nucleic Acids Symposium, Series 8*, 1980, p 33.
47. For examples, see: (a) Curran, W. V.; Angier, R. B. *J. Org. Chem.* **1966**, *31*, 201. (b) Ueda, T.; Tanaka, H. *Chem. Pharm. Bull.* **1970**, *18*, 1491. (c) Niedballa, U.; Vorbrüggen, H. *J. Org. Chem.* **1974**, *39*, 3660.
48. (a) Inoue, H.; Ueda, T. *Chem. Pharm. Bull.* **1971**, *19*, 1743. (b) Holy, A. *Coll. Czech. Chem. Commun.* **1975**, *40*, 738. (c) Matsuda, A.; Inoue, H.; Ueda, T. *Chem. Pharm. Bull.* **1978**, *26*, 2340. (d) Inoue, H.; Ueda, T. *Chem. Pharm. Bull.* **1978**, *26*, 2657.
49. Inoue, H.; Tomita, S.; Ueda, T. *Chem. Pharm. Bull.* **1975**, *23*, 2614.
50. Other examples for the synthesis of 6-substituted derivatives: (a) Fourrey, J.-L.; Henry, G.; Jouin, P. *Tetrahedron Lett.* **1979**, 951. (b) Rosenthal, A.; Dodd, R. H. *Carbohydr. Res.* **1980**, *85*, 15.
51. (a) Tanaka, H.; Hayakawa, H.; Miyasaka, T. *Chem. Pharm. Bull.* **1981**, *29*, 3565. (b) Tanaka, H.; Hayakawa, H.; Miyasaka, T. *Tetrahedron* **1982**, *38*, 2635.
52. Synthesis of 6-trimethylsilyluridine based on the LDA-lithiation of **65** has been reported: Ikehira, H.; Matsuura, T.; Saito, I. *Tetrahedron Lett.* **1984**, *25*, 3325.
53. Tanaka, H.; Hayakawa, H.; Miyasaka, T. Unpublished results.
54. (a) Ikehira, H.; Matsuura, T.; Saito, I. *Tetrahedron Lett.* **1985**, *26*, 1743. (b) Satoh, K.; Tanaka, H.; Andoh, A.; Miyasaka, T. *Nucleosides Nucleotides* **1986**, *5*, 461.
55. Tanaka, H.; Haraguchi, K.; Koizumi, Y.; Fukui, M.; Miyasaka, T. *Can J. Chem.* **1986**, *64*, 1560.

56. Tanaka, H.; Iijima, S.; Matsuda, A.; Hayakawa, H.; Miyasaka, T.; Ueda, T. *Chem. Pharm. Bull.* **1983**, *31*, 1222.

57. (a) Tanaka, H.; Hayakawa, H.; Haraguchi, K.; Miyasaka, T. *Nucleosides Nucleotides* **1985**, *4*, 607. (b) Miyasaka, T.; Tanaka, H.; Satoh, K.; Imahashi, M.; Yamaguchi, K.; Iitaka, Y. *J. Heterocycl. Chem.* **1987**, *24*, 873.

58. (a) Kapuler, A. M.; Monny, C.; Michelson, A. M. *Biochim. Biophys. Acta* **1970**, *217*, 18. (b) Schweizer, M. P.; Witkowski, J. T.; Robins, R. K. *J. Am. Chem. Soc.* **1971**, *93*, 277.

59. Tanaka, H.; Matsuda, A.; Iijima, S.; Hayakawa, H.; Miyasaka, T. *Chem. Pharm. Bull.* **1983**, *31*, 2164.

60. (a) Tanaka, H.; Hayakawa, H.; Obi, K.; Miyasaka, T. *Tetrahedron Lett.* **1985**, *26*, 6229. (b) Tanaka, H.; Hayakawa, H.; Obi, K.; Miyasaka, T. *Tetrahedron* **1986**, *42*, 4187.

61. After our preliminary communication (ref. 60a) was published, homolytic stannylation of vinylsulfones was reported, and an electron transfer mechanism was proposed for the reaction: Watanabe, Y.; Ueno, Y.; Araki, T.; Endo, T.; Okawara, M. *Tetrahedron Lett.* **1986**, *27*, 215.

62. Tanaka, H.; Hayakawa, H.; Iijima, S.; Haraguchi, K.; Miyasaka, T. *Tetrahedron* **1985**, *41*, 861.

63. Hayakawa, H.; Tanaka, H.; Maruyama, Y.; Miyasaka, T. *Chem. Lett.* **1985**, 1401.

64. Rabi, J. A.; Fox, J. J. *J. Am. Chem. Soc.* **1973**, *95*, 1628.

65. Hayakawa, H.; Tanaka, H.; Obi, K.; Itoh, M.; Miyasaka, T. *Tetrahedron Lett.* **1987**, *28*, 87.

66. For applications of this method, see: (a) Nawrot, B.; Malkiewicz, A. *Nucleosides Nucleotides* **1989**, *8*, 1499. (b) Armstrong, R. W.; Gupta, S.; Whelihan, F. *Tetrahedron Lett.* **1989**, *30*, 2057.

67. For a review, see: Chu, C. K.; Cutler, S. J. *J. Heterocycl. Chem.* **1986**, *23*, 289.

68. For a bibliography, see: Remy, R. J.; Secrist III, J. A. *Nucleosides Nucleotides* **1985**, *4*, 411.

69. Miyasaka, T.; Tanaka, H.; Baba, M.; Hayakawa, H.; Walker, R. T.; Balzarini, J.; De Clercq, E. *J. Med. Chem.* **1989**, *32*, 2507.

70. Baba, M.; Tanaka, H.; De Clercq, E.; Pauwels, R.; Balzarini, J.; Schols, D.; Nakashima, H.; Perno, C.-F.; Walker, R. T.; Miyasaka, T. *Biochem. Biophys. Res. Commun.* **1989**, *165*, 1375.

71. Fyfe, J. A.; Keller, P. M.; Furman, P. A.; Miller, R. L.; Elion, G. B. *J. Biol. Chem.* **1978**, *253*, 8721.

72. For reviews, see: De Clercq, E. *Med. Res. Rev.* **1993**, *13*, 229 and *Med. Res. Rev.* **1996**, *16*, 125.

73. Pauwels, R.; Andries, K.; Desmyter, J.; Schols, D.; Kukla, M. J.; Breslin, H. J.; Raeymaeckers, A.; Gelder, J. V.; Woestenborghs, R.; Heykants, J.; Schellekens, K.; Janssen, M. A. C.; De Clercq, E.; Janssen, P. A. J. *Nature* **1990**, *343*, 470.

74. Merluzzi, V. J.; Hargrave, K. D.; Labadia, M.; Grozinger, K.; Skoog, M.; Wu, J. C.; Shin, C.-K.; Eckner, K.; Hattox, S.; Adams, J.; Rosenthal, A. S.; Faanes, R.; Eckner, R. J.; Koup, R. A.; Sullivan, J. L. *Science* **1990**, *250*, 1411.

75. Goldman, M. E.; Nunberg, J. H.; O'Brien, J. A.; Quintero, J. C.; Schleif, W. A.; Freund, K. F.; Gaul, S. L.; Saari, W. S.; Wai, J. S.; Hoffman, J. M.; Anderson, P. S.; Hupe, D. J.; Emini, E. A.; Stern, A. M. *Proc. Natl. Acad. Sci. USA* **1991**, *88*, 6863.

76. (a) Camarasa, M.-J.; Pérez-Pérez, M.-J.; San-Félix, A.; Balzarini, J.; De Clercq, E. *J. Med. Chem.* **1992**, *35*, 2721. (b) Balzarini, J.; Pérez-Pérez, M.-J.; San-Félix, A.; Schols, D.; Perno, C.-F.; Vandamme, A.-M.; Camarasa, M.-J.; De Clercq, E. *Proc. Natl. Acad. Sci. USA* **1992**, *89*, 4392.

77. Pauwels, R.; Andries, K.; Debyser, Z.; Van Daele, P.; Schols, D.; Stoffels, P.; De Vreese, K.; Woestenborghs, R.; Vandamme, A.-M.; Janssen, C. G. M.; Anne, J.; Cauwenbergh, G.; Desmyter, J.; Heykants, J.; Janssen, M. A. C.; De Clercq, E.; Janssen, P. A. J. *Proc. Natl. Acad. Sci. USA* **1993**, *90*, 1711.

78. Zhang, H.; Vrang, L.; Bäckbro, K.; Lind, P.; Sahlberg, C.; Unge, T.; Öberg, B. *Antiviral Res.* **1995**, *28*, 331.

79. Young, S. D.; Britcher, S. F.; Tran, L. O.; Payne, L. S.; Lumma, W. C.; Lyle, T. A.; Huff, J. R.; Anderson, P. S.; Olsen, D. B.; Carroll, S. S.; Pettibone, D. J.; Obrien, J. A.; Ball, R. G.; Balani, S. K.; Lin, J. H.; Chen, I.-W.; Schleif, W. A.; Sardana, V. V.; Long, W. J.; Byrnes, V. W.; Emini, E. A. *Antimicrob. Agents Chemother.* **1995**, *39*, 2602.

80. Hanasaki, Y.; Watanabe, H.; Katsuura, K.; Takayama, H.; Shirakawa, S.; Yamaguchi, K.; Sasaki, S.; Ijichi, K.; Fujiwara, M.; Konno, K.; Yokota, T.; Shigeta, S.; Baba, M. *J. Med. Chem.* **1995**, *38*, 2038.

81. Tanaka, H.; Baba, M.; Hayakawa, H.; Sakamaki, T.; Miyasaka, T.; Ubasawa, M.; Takashima, H.; Sekiya, K.; Nitta, I.; Shigeta, S.; Walker, R. T.; Balzarini, J.; De Clercq, E. *J. Med. Chem.* **1991**, *34*, 349.

82. Tanaka, H.; Baba, M.; Ubasawa, M.; Takashima, H.; Sekiya, K.; Nitta, I.; Shigeta, S.; Walker, R. T.; De Clercq, E.; Miyasaka, T. *J. Med. Chem.* **1991**, *34*, 1394.

83. Tanaka, H.; Baba, M.; Saito, S.; Miyasaka, T.; Takashima, H.; Sekiya, K.; Ubasawa, M.; Nitta, I.; Walker, R. T.; Nakashima, H.; De Clercq, E. *J. Med. Chem.* **1991**, *34*, 1508.

84. Tanaka, H.; Takashima, H.; Ubasawa, M.; Sekiya, K.; Nitta, I.; Baba, M.; Shigeta, S.; Walker, R. T.; De Clercq, E.; Miyasaka, T. *J. Med. Chem.* **1992**, *35*, 337.

85. Tanaka, H.; Miyasaka, T.; Sekiya, K.; Takashima, H.; Ubasawa, M.; Nitta, I.; Baba, M.; Walker, R. T.; De Clercq, E. *Nucleosides Nucleotides* **1992**, *11*, 447.

86. Tanaka, H.; Takashima, H.; Ubasawa, M.; Sekiya, K.; Nitta, I.; Baba, M.; Shigeta, S.; Walker, R. T.; De Clercq, E.; Miyasaka, T. *J. Med. Chem.* **1992**, *35*, 4713.

87. Tanaka, H.; Baba, M.; Yamamoto, T.; Mori, S.; Walker, R. T.; De Clercq, E.; Miyasaka, T. *Bioorg. Med. Chem. Lett.* **1993**, *3*, 1681.

88. Tanaka, H.; Baba, M.; Takahashi, E.; Matsumoto, K.; Kittaka, A.; Walker, R. T.; De Clercq, E.; Miyasaka, T. *Nucleosides Nucleotides* **1994**, *13*, 155.

89. Tanaka, H.; Takashima, H.; Ubasawa, M.; Sekiya, K.; Inouye, N.; Baba, M.; Shigeta, S.; Walker, R. T.; De Clercq, E.; Miyasaka, T. *J. Med. Chem.* **1995**, *38*, 2860.

90. Baba, M.; De Clercq, E.; Iida, S.; Tanaka, H.; Nitta, I.; Ubasawa, M.; Takashima, H.; Sekiya, K.; Umezu, K.; Nakashima, H.; Shigeta, S.; Walker, R. T.; Miyasaka, T. *Antimicrob. Agents Chemother.* **1990**, *34*, 2358.

91. Baba, M; De Clercq, E.; Tanaka, H.; Ubasawa, M.; Takashima, H.; Sekiya, K.; Nitta, I.; Umezu, K.; Nakashima, H.; Mori, S.; Shigeta, S.; Walker, R. T.; Miyasaka, T. *Proc. Natl. Acad. Sci. USA* **1991**, *88*, 2356.

92. Baba, M.; De Clercq, E.; Tanaka, H.; Ubasawa, M.; Takashima, H.; Sekiya, K.; Nitta, I.; Umezu, K.; Walker, R. T.; Mori, S.; Ito, M.; Shigeta, S.; Miyasaka, T. *Mol. Pharmacol.* **1991**, *39*, 805.

93. The preparation of 6-arylthio- and 6-arylselenoacylouridines based on the LDA lithiation has been reported: Pan, B.-C.; Chen, Z.-H.; Piras, G.; Dutschman, G. E.; Rowe, E. C.; Cheng, Y.-C.; Chu, S.-H. *J. Heterocyclic Chem.* **1994**, *31*, 177.

94. Addition-elimination reactions of 6-methylsulfonyluracil derivatives have been reported: Kim, D.-K.; Kim, Y.-W.; Gam, J.; Lim, J.; Kim, K. H. *Tetrahedron Lett.* **1995**, *36*, 6257.

95. Preparation of the 5-nitro derivatives of HEPT has been carried out by an addition-elimination reaction: Benhida, R.; Aubertin, A.-M.; Grierson, D. S.; Monneret, C. *Tetrahedron Lett.* **1996**, *37*, 1031.

96. Deoxy analogs of HEPT, which also lack the ether oxygen in the acyclic structure, have been synthesized by a palladium-catalyzed coupling reaction: Pontikis, R.; Monneret, C. *Tetrahedron Lett.* **1994**, *35*, 4351.

97. For a QSAR study of HEPT analogs, see: Hansch, C.; Zhang, L. *Bioorg. Med. Chem. Lett.* **1992**, *2*, 1165.

98. Baba, M.; Yuasa, S.; Niwa, T.; Yamamoto, M.; Yabuuchi, S.; Takashima, H.; Ubasawa, M.; Tanaka, H.; Miyasaka, T.; Walker, R. T.; Balzarini, J.; De Clercq, E.; Shigeta, S. *Biochem. Pharmacol.* **1993**, *45*, 2507.

99. (a) Baba, M.; Shigeta, S.; Yuasa, S.; Takashima, H.; Sekiya, K.; Ubasawa, M.; Tanaka, H.; Miyasaka, T.; Walker, R. T.; De Clercq, E. *Antimicrob. Agents Chemother.* **1994**, *38*, 688. (b) Baba, M.; Tanaka, H.; Miyasaka, T.; Yuasa, S.; Ubasawa, M.; Walker, R. T.; De Clercq, E. *Nucleosides Nucleotides* **1995**, *14*, 575. (c) Brennan, T. M.; Taylor, D. L.; Bridges, C. G.; Leyda, J. P.; Tyms, A. S. *Antiviral Res.* **1995**, *26*, 173.

100. Baba, M.; Ito, M.; Shigeta, S.; Tanaka, H.; Miyasaka, T.; Ubasawa, M.; Umezu, K.; Walker, R. T.; De Clercq, E. *Antimicrob. Agents Chemother.* **1991**, *35*, 1430.

101. Yuasa, S.; Sadakata, Y.; Takashima, H.; Sekiya, K.; Inouye, N.; Ubasawa, M.; Baba, M. *Mol. Pharmacol.* **1993**, *44*, 895.

102. Kohlstaedt, L. A.; Wang, J.; Friedman, J. M.; Rice, P. A.; Steitz, T. A. *Science* **1992**, *256*, 1783.

103. Smerdon, S. J.; Jäger, J.; Wang, J.; Kohlstaedt, L. A.; Chirino, A. J.; Friedman, J. M.; Rice, P. A.; Steitz, T. A. *Proc. Natl. Acad. Sci. USA* **1994**, *91*, 3911.

104. Ren, J. S.; Esnouf, R.; Garman, E.; Jones, Y.; Somers, D.; Ross, C.; Kirby, I.; Keeling, J.; Darby, G.; Stuart, D.; Stammers, D. *Nature Struct. Biol.* **1995**, *2*, 293.

105. Ding, J.; Das, K.; Tantillo, C.; Zhang, W.; Clark, A. D. J.; Jessen, S.; Lu, X.; Hsiou, Y.; Jacob-Molina, A.; Andries, K.; Pauwels, R.; Moereels, H.; Koymans, L.; Janssen, P. A. J.; Smith, R. H. J.; Kroeger Koepke, R.; Michejda, C. J.; Hughes, S. H.; Arnold, E. *Structure* **1995**, *3*, 365.

106. Ding, J.; Das, K.; Moereels, H.; Koymans, L.; Andries, K.; Janssen, P. A. J.; Hughes, S. H.; Arnold, E. *Nature Struct. Biol.* **1995**, *2*, 407.

107. Ren, J.; Esnouf, R.; Hopkins, A.; Ross, C.; Jones, Y.; Stammers, D.; Stuart, D. *Structure* **1995**, *3*, 915.

108. Kraulis, P. J. *J. Appl. Crystallogr.* **1991**, *24*, 946.

109. Merritt, E. A.; Murphy, M. E. P. *Acta Crystallogr.* **1994**, *D50*, 869.

110. Frank, K. B.; Noll, G. J.; Connell, E. V.; Sim, I. S. *J. Biol. Chem.* **1991**, *266*, 14232.

111. Althaus, I. W.; Chou, J. J.; Gonzales, A. J.; Deibel, M. R.; Chou, K.-C.; Kezdy, F. J.; Romero, D. L.; Aristoff, P. A.; Tarpley, W. G.; Reusser, F. *J. Biol. Chem.* **1993**, *268*, 6119.

112. Spence, R. A.; Kati, W. M.; Anderson, K. S.; Johnson, K. A. *Science* **1995**, *267*, 988.

113. Hopkins, A. L.; Ren, J.; Esnouf, R. M.; Willcox, B. E.; Jones, E. Y.; Ross, C.; Miyasaka, T.; Walker, R. T.; Tanaka, H.; Stammers, D. K.; Stuart, D. I. *J. Med. Chem.* **1996**, *39*, 1589.

114. Balzarini, J.; Karlsson, A.; De Clercq, E. *Mol. Pharmacol.* **1993**, *44*, 694.

115. Seki, M.; Sadakata, Y.; Yuasa, S.; Baba, M. *Antiviral Chem. Chemother.* **1995**, *6*, 73.

SIALIDASE INHIBITORS AS ANTI-INFLUENZA DRUGS

Mark von Itzstein and Jeff C. Dyason

I. GENERAL INTRODUCTION

A recent outbreak of a "bird flu" in Hong Kong late in 1997 and the likelihood that the disease was spreading through contact between people increased the prospect of a pandemic that could cause the deaths of thousands of individuals. The World Health Organization (WHO) considers influenza a serious socioeconomic disease and

Advances in Antiviral Drug Design
Volume 3, pages 145–160.
Copyright © 1999 by JAI Press Inc.
All rights of reproduction in any form reserved.
ISBN: 0-7623-0201-1

therefore it closely monitors the appearance and impact of the virus on a worldwide basis. Already during this century outbreaks of influenza virus infection have reached the proportion of pandemics resulting in the deaths of millions of humans. For example, at the close of World War I, the Spanish influenza pandemic claimed the lives, either directly or indirectly, of at least 20 million people.[1] The most vulnerable in the human population appear to be the immunocompromised such as the elderly, the very young and those with existing conditions such as chronic pulmonary and heart disease.

The identification of the causative agent of influenza was not made until 1933, when it was determined to be a virus.[2] These viruses are members of the orthomyxoviridae family and are further classified, on serological differences, into three distinct types, A, B, and C. Generally it is thought that only types A and B are of major concern to humans with both types A and B being responsible for severe widespread epidemics, while type A also appears to cause pandemics. As described in a review by Herrler et al.,[3] influenza A virus appears to have a high rate of mutation in its surface glycoproteins as a result of both antibody pressure and genetic reassortment through the mixing of nonhuman and human strains. From these mechanisms of mutation different strains of the virus have been classified on the basis of the hemagglutinin (e.g. subtypes H1, H2, H3) and sialidase (e.g. subtypes N1, N2, N9) cell-surface glycoprotein antigens. For example the most recent outbreak of influenza in Hong Kong the virus strain has been determined to be H5N1 and was first isolated in poultry.

Given the fact that the virus does mutate at a fairly rapid rate, vaccines raised against previous strains don't necessarily afford protection for a new season of influenza. Some excellent reviews describe present-day drug therapies.[4-6] For example, compounds such as amantadine (1) and rimantadine (2), which are specific anti-influenza A drugs,[7] are thought to block ion channels of the viral M2 protein, subsequently interfering with virus uncoating and thereby inhibiting viral replication.[4,5] Another example of a clinically useful drug is the synthetic nucleoside ribavirin (3) which appears to be a broad-spectrum antiviral agent with good efficacy in the treatment of respiratory infections, including influenza (see for example ref. 5). Unfortunately, all of these compounds suffer toxicity problems and do not provide long term solutions for the treatment of influenza.

(1) R = NH$_2$

(2) R = CH(Me)NH$_2$

(3)

The virus cell surface glycoproteins, hemagglutinin and sialidase, are both sialic acid-recognizing proteins and appear as spikes that project from the viral cell surface.[3] These glycoproteins are considered by many researchers as drug design targets and may play a crucial role in the discovery of exciting therapeutic agents against the virus.[4,5,8–10]

The role of sialic acids in influenza virus infection has been reviewed recently by Herrler et al.[3] Efforts towards the design and synthesis of potential anti-influenza drugs based on inhibition of the influenza virus hemagglutinin[4,8,11–13] and sialidase[5,6,8,9,12–14] glycoproteins have also been reviewed. The present contribution will restrict itself to a discussion on the design and synthesis of influenza virus sialidase inhibitors as potential anti-influenza drugs.

II. INFLUENZA VIRUS SIALIDASE

A. Overview

Influenza virus sialidase appears to be important in the replication cycle of the virus and plays a number of important roles.[3,15] One of these roles appears to be in the facilitation of virus movement through the respiratory tract mucus. Influenza virus sialidase is a tetrameric glycoprotein with a subunit molecular weight of 60 kDa. The enzyme acts as a glycohydrolase cleaving α-ketosidically linked terminal *N*-acetyl-D-neuraminic acid (Neu5Ac **4**) residues (Scheme 1) on the mucosal sialoglycoproteins reducing the mucus viscosity, and concomitantly reducing the concentration of highly sialylated mucosal secretions which may otherwise significantly slow the virus and its bid to be successful in infection.[3,15] As a result of the enzyme's action,

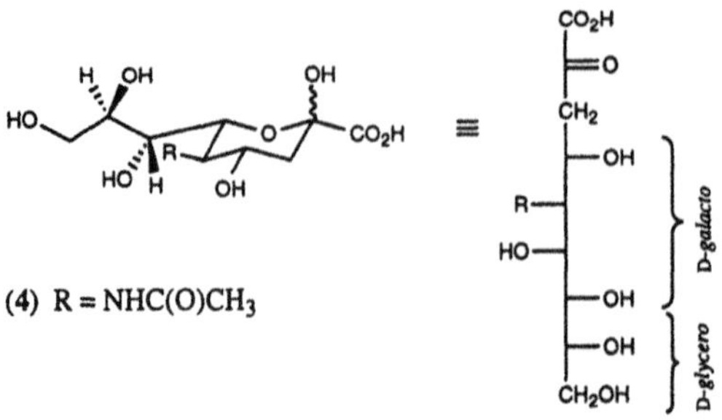

(4) R = NHC(O)CH$_3$

viral penetration to the target epithelial cells is most likely enhanced. Another role that has been associated with this viral enzyme is its capacity to assist in the release of the newly synthesized virions from the infected cells.[16,17] It has been the opinion of the drug designers that inhibition of the sialidase action may permit the immune system time to mount an appropriate response to the newly synthesized progeny which are clumped at the infected cell's surface before they are released.

As a result of antigenic variation in influenza A virus sialidase, nine subtypes are now known with two of these subtypes (N2 and N9) being from strains that have been isolated from humans.[15,18] It is interesting to note that approximately 50% sequence homology between these subtypes has been found.[19] Structural information (Figure 1) for influenza A N2[19,20] and N9[21,22] sialidases, and influenza B virus sialidase[23] has been reported and the active site, identified by comparison of protein sequences from several influenza N2 sialidases, appears to be located in a cavity on the upper surface of the head domain of the glycoprotein.[24] One of the most interesting observations from these studies was that 18 of the residues in the active are conserved in all

Scheme 1.

Figure 1. An influenza virus sialidase monomer unit.

known influenza A and B virus sialidase sequences characterized to date,[14,24] of which 15 of the conserved residues are charged. This degree of charge is quite unusual as most glycohydrolases are generally thought to have only a small number, two or three, charged residues. Perhaps not surprisingly, eight of the strain-invariant active site residues are positioned to make direct contact with *N*-acetyl-D-neuraminic acid (**4**) and its unsaturated glycal 2-deoxy-2,3-didehydro-*N*-acetyl-D-neuraminic acid (Neu5Ac2en **5**) bound into the catalytic site, with a further tier of ten residues which appears to establish a structural scaffold for the catalytic site.[25] Over the past decade a number of crystal structures of complexes between influenza A siali-

(5) R = NHAc

dase and substrate-like compounds or inhibitors,[26-31] and between influenza B sialidase and Neu5Ac2en[32] have also been determined.

The action of the influenza virus sialidase at a molecular level has been investigated by a number of methods. These investigations have included mutation studies,[33] molecular modeling studies of the active site,[34,35] biochemical studies,[34,36] and NMR studies of the enzyme reaction.[34] We have proposed[34-36] that a stabilized sialosyl cation transition-state intermediate is involved when influenza virus sialidase catalyzes the release of Neu5Ac from α-sialosides.

B. Inhibitors of Influenza Virus Sialidase

A number of reviews have summarized various classes of compounds that have been proposed as inhibitors of influenza virus sialidase.[5,8,37] Only some of the more recent developments will be described in this contribution.

Carbohydrate-Based Inhibitors of Influenza Virus Sialidase

A range of N-acetyl-D-neuraminic acid derivatives have shown varying degrees of inhibition of influenza virus sialidase and these have been recently reviewed [see for example ref. 8 and references therein]. It has been found that Neu5Ac (4) itself inhibits influenza A virus sialidase with an apparent K_i of approximately 10^{-3} M.[16] Although given that it is only the α-anomer that binds to the enzyme's active site and that the β-anomer is significantly more thermodynamically stable, then the true inhibition constant of the α-anomer would be closer to 10^{-4} M.

Noteworthy is the observation that α-ketosidically-linked S-[38-40] and N-glycosides[38] of Neu5Ac appear to be sialidase resistant. Furthermore some simple alkyl and aromatic (e.g. 6 and 7) α-glycosides have also shown inhibition of influenza A sialidase.[38] Also incorpora-

(6) R = NHAc, X = S
(7) R = NHAc, X = N

tion of sulfur-linked Neu5Ac into ganglioside structures such as **8**[41] afforded the GM_3 ganglioside resistance to influenza virus sialidase.[39,41]

While these glycosides are reasonable inhibitors of influenza virus sialidase, the most potent inhibitors described to date are those based on the unsaturated glycal of Neu5Ac, compound (**5**). Glycal **5**, 2-deoxy-2,3-didehydro-*N*-acetylneuraminic acid (Neu5Ac2en, **5**), was first reported as an inhibitor of sialidases almost three decades ago.[42] Subsequently, a range of C5 substituted derivatives were evaluated against sialidases from *V. cholerae*, influenza A and B viruses, and Newcastle disease virus.[43] Although early work indicated that the putative transition state analog **5** and one of its derivatives inhibited influenza virus replication in tissue culture studies, these compounds failed to provide any protection against influenza infection when mice were dosed either by an intranasal or intravenous route.[44] It was suggested in another study that this may be the result of rapid excretion of the compound from the body.[45]

With a significant amount of structural information[19,20,24,26] available for influenza virus sialidase the opportunity for the design of potent inhibitors of the enzyme using computer-assisted manual examination and molecular modeling studies has been exploited.[34,35,46,47]

(8)   R = NHAc

Initial computer assisted visual inspection of the active site structure of sialidase complexed with both Neu5Ac and Neu5Ac2en, has provided valuable information on the method of binding and the important interactions between the carbohydrate and the enzyme. One of the most important interactions appears to be that between the carboxylate of both of these sialic acids with the so-called triarginyl cluster (Arg118, Arg292, and Arg371).[35] This interaction appears to be important both enzymatically, on binding the sugar ring is flattened out prior to the cleavage of the glycosidic bond, and structurally with the electrostatic interaction between the negatively charged carboxylate and the positively charged amine groups of the arginines being a major contributor to the binding of substrates and substrate-based inhibitors.[34,46] These findings further support the notion that Neu5Ac2en is a reasonable transition state analog and provides a plausible explanation for the compound's apparent higher affinity for the enzyme in comparison with N-acetyl-D-neuraminic acid ($K_i = 1 \times 10^{-3}$ M).

The program GRID[48] was then used in conjunction with the X-ray crystal structure of sialidase to determine whether the Neu5Ac2en template could be improved upon to produce a better inhibitor. Essentially GRID uses the crystal structure to evaluate the electrostatic and steric interactions between various probes at points on a regular three-dimensional grid which is overlaid on the active site. These probes represent common organic functional groups (e.g. carboxyl group, amino group, etc.) and provide information as to the best position and orientation for each probe within the active site. The best way of viewing the results involved the overlay of the three-dimensional contour map for each probe over the active site of the sialidase, which also had present a solvent accessible surface for the active site residues.

In an attempt to validate the method the carboxyl probe was used in the first instance as it was thought that if the program was reliable then it should predict the interaction between the N-acetyl-D-neuraminic acid carboxylate and the triarginyl cluster. Indeed, this was predicted by GRID to have an interaction energy of –9 kcal/mol and moreover the volume of the interaction was directly superimposable with the carboxyl group of N-acetyl-D-neuraminic acid.[47] Interestingly, no other significant interactions were predicted for this probe by GRID. This was quite promising and suggested that the method would be useful in a predictive way for other probes. The other probes of interest were the amino probe, the hydroxyl probe, and the methyl probe.[47] The amino probe showed three sites of interest with very favorable interaction energies (16–18 kcal/mol). Site 1 was found to be adjacent to the 4-hydroxyl group position of *N*-acetyl-D-neuraminic acid, site 2 was located beneath the C4 position of the sugar ring and site 3 was on the floor of the active site between the glycerol side chain and 5-*N*-acetyl group. The hydroxyl probe showed similar interactions to the carboxylate probe around the triarginyl cluster, but at a lower level (–8 kcal/mol). This probe also showed small areas of interaction along the glycerol side chain at –6 kcal/mol, but surprisingly there was no interaction at this level predicted for the 4-hydroxyl group of *N*-acetyl-D-neuraminic acid. The methyl probe showed only one main area of interaction at a contour level of –4 kcal/mol and this was adjacent to the methyl group of the 5-*N*-acetyl group of *N*-acetyl-D-neuraminic acid. A full analysis of the main probe interactions has been described elsewhere.[47]

The GRID results while being extremely helpful in the initial design process really only gave qualitative answers as to which modifications on the basic template would provide the best interactions. In an effort to provide more quantitative data on whether one compound would be a better inhibitor than another, a molecular dynamics/energy minimization protocol was developed to analyze the structural and energetic effects on functional group modification of the C4 position of the basic template Neu5Ac2en.[49]

From these above-described studies a series of 4-substituted-4-de-oxy-*N*-acetyl-D-neuraminic acid derivatives were prepared and biologically evaluated.[46,50–52] The synthesis of a 4-amino substituted Neu5Ac2en derivative (**9**) was achieved from the readily available Neu5Ac2en itself.[51,52] Evaluation against influenza virus sialidase indicated an inhibition constant (K_i) of 4×10^{-8} M, which represented an increase of two orders of magnitude in inhibition when compared with the base template Neu5Ac2en ($K_i \; 4 \times 10^{-6}$ M).[50] This observed increase is in keeping with the findings of a quantitative study[49] of C4 modified sialic acids. Further modeling studies led to the notion that replacement of the amino group with the larger more basic guanidino functionality (**10**)[47] should produce an even more potent inhibitor. The target compound **10** was prepared from the 4-amino derivative **9**[51,52]

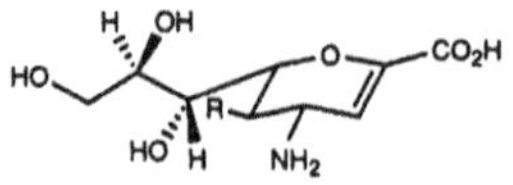

(9)   R = NHAc

and was indeed found to be a more potent inhibitor (at steady state K_i = 3 × 10^{-11} M).[53] A number of molecular modeling studies clearly showed that the guanidino substituent was within hydrogen bond distance with two conserved acid residues in the C4 binding pocket.[35,46,47] Subsequently, crystallographic studies of these inhibitors soaked into influenza A virus sialidase crystals supported the predicted mode of binding for the C4 substituents.[46]

Both 4-amino-4-deoxy- and 4-deoxy-4-guanidino-Neu5Ac2en compounds, 9 and 10, respectively, were found to selectively inhibit influenza virus sialidase, while other sialidases from both bacterial and mammalian sources were found to be resistant.[50] This selectivity was thought to be a result of both steric and electronic effects within the active sites of the noninfluenza virus sialidases.[50] Subsequently with the elucidation of the structures of both *Vibrio cholerae*[54] and *Salmonella typhimurium*[55] sialidases completed, support for the proposed architectural differences around the critical C4 binding pocket was indeed found.[54]

A range of biological studies have found that both 4-amino-4-deoxy- and 4-deoxy-4-guanidino-Neu5Ac2en effectively inhibit both the replication of influenza A and B viruses *in vitro*[46,56] and *in vivo* when they were administered via an intranasal route in small animals.[46,57,58] In 1994 4-deoxy-4-guanidino-Neu5Ac2en (10) (Zanamivir) was taken into clinical trials and was found to be both safe and effective in the prophylaxis and early therapy of influenza

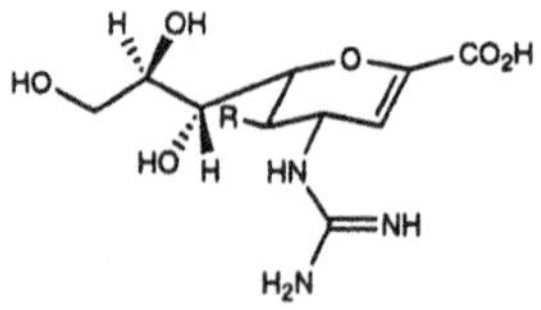

(10) R = NHAc

infection in humans.[59] More recently (1997/1998), phase three clinical trials of the compound (Zanamivir) have commenced and data is now awaited before the compound, if the data is sufficiently positive, can be lodged for formal registration.

The success of these *N*-acetyl-D-neuraminic acid derivatives as effective anti-influenza agents promoted significant activity for the discovery of improved inhibitors. This was thought of particular importance because of the lack of oral bioavailability of the most potent inhibitors. A significant range of *N*-acetyl-D-neuraminic acid derivatives have been prepared and these have been well reviewed in recent literature.[8,9] None of these more recently prepared derivatives have yet progressed to clinical trials.

Non-Carbohydrate-Based Sialylmimetics as Inhibitors of Influenza Virus Sialidase

Over the past two or 3 years the preparation of next generation non-carbohydrate-based sialylmimetics of 4-amino-4-deoxy- and 4-deoxy-4-guanidino-Neu5Ac2en has been investigated. Two quite distinct approaches have been taken towards the synthesis of these sialylmimetics. The first of these led to a series of compounds that were designed based on 4-(*N*-acetylamino)benzoic acid, which used the benzene ring as a substitute for the dihydropyran ring of the sialic acid.[28,29,60,61] The rationale behind the choice of the benzene ring template was that it was thought to provide the appropriate spacing for the various functional groups compared to Neu5Ac2en, and significantly reduced the number of stereochemical centers associated with the *N*-acetyl-D-neuraminic acid derivatives.[28] Of course the assumption made was that the additional electronics and planarity associated with a benzene ring would have minimal impact (or indeed positive influences) on binding of these sialylmimetics. While some of these compounds have demonstrated reasonable affinity for influenza virus sialidase, for example compound **11** had IC_{50} values of 10^{-6} M, none have been found to be active *in vivo*.[29,61]

More recently, a second approach towards the synthesis of non-carbohydrate sialylmimetics has been reported.[31,62] The precursors to these carbocyclic mimetics (**12** and **13**) of 4-amino-4-deoxy-Neu5Ac2en were quinic and shikimic acids and through synthetic

(11)

manipulation have the correct stereochemistry when compared with
N-acetyl-D-neuraminic acid with the exception that the glycerol side
chain is replaced by an alcohol or ether functionality.[31] The placement
of the double bond in the cyclohexene ring was found to be critical
and had a dramatic influence on their inhibition of influenza A
sialidase activity;[31] for example compound **13** was found to have an
IC_{50} of 6.3×10^{-6} M, whereas no inhibitory activity was observed for
compound **12** at concentrations up to 2×10^{-4} M. Further elaboration
of compound **13** by the introduction of lipophilic substituents on the
C3 hydroxyl group provided a series of compounds that were significantly more potent than the parent compound.[31] It is thought[31] that the
introduction of these aliphatic moieties captures some hydrophobic
interactions in the glycerol side-chain binding pocket of the enzyme.
Compound **14** which bears a 3-pentyloxy substituent is to date the
most potent compound of this series and its potency was found to
compare to that of the initial binding mode of 4-deoxy-4-guanidino-
Neu5Ac2en (**10**) for both enzyme inhibition and plaque assays.[31]
Good bioavailability in animal models, and oral efficacy in the treatment of influenza virus infection in small animals, has been observed
with a prodrug form, also known as GS4104, of compound **14**.[31] This
compound has now entered phase 2 clinical trials.[31]

(12)

(13) R = H
(14) R = CH(CH$_2$CH$_3$)$_2$

The Emergence of Resistant Mutants to Sialidase-Based Influenza Drug Candidates

Clearly there is the possibility of the emergence of drug-resistant mutants when the virus is placed under pressure from the drug. The obvious site for significant mutation would be in the active site of the sialidase glycoprotein where the drug candidates act, although it is also possible that the virus could alter its hemagglutinin glycoprotein to modulate its requirement for N-acetyl-D-neuraminic acid (**4**) in the adhesion event. A number of drug-resistant variants have been observed when influenza A virus strains were cultured in the presence of 4-deoxy-4-guanidino-Neu5Ac2en (**10**), of which a number had a reduced sensitivity to the drug compared with the parent virus.[63–66] The fact that drug-resistant mutants can be produced *in vitro* is not surprising; however the significance of this observation in the clinic is not yet known.

III. CONCLUSION

With a number of sialidase-based influenza candidate drugs presently in clinical trial, the prospect of a cure for influenza is real. It is indeed exciting times and the influenza story has been a shining example of structure-based drug design.

NOTE ADDED IN PROOF

Status of Anti-Influenza Drugs

After successful completion of phase II and III clinial trials,[67] Zanamivir (market name Relenza®) is now available on the market for the treatment of influenza in Australia and Europe.

The results of phase II clinical trials for the orally bioavailable prodrug, GS4104, are now available.[68] Phase II trials were commenced at the beginning of 1999 and in May of this year a New Drug Application was filed with the U.S. Food and Drug Administration (FDA).

REFERENCES

1. Crosby, A. W. *America's Forgotten Pandemic*; Cambridge University Press: Cambridge, 1989.
2. Smith, W. C.; Andrews, C. H.; Laidlaw, P. P. *Lancet* **1933**, *2*, 66.
3. Herrler, G.; Hausmann, J.; Klenk, H.-D. In *Biology of the Sialic Acids*; Rosenberg, A., Ed.; Plenum Press: New York, 1995, p 315.
4. Meanwell, N. A.; Krystal, M. *Drug Discovery Today* **1996**, *1*, 316.
5. Meanwell, N. A.; Krystal, M. *Drug Discovery Today* **1996**, *1*, 388.
6. Bamford, M. J. *J. Enzy. Inhib.* **1995**, *10*, 1.
7. Barrish, J. C.; Zahler, R. In *Ann. Rep. Med. Chem.*; Bristol, J. A.; Plattner, J. J., Ed.; Academic Press: San Diego, 1993, p 131.
8. von Itzstein, M.; Thomson, R. J. *Curr. Med. Chem.* **1997**, *4*, 185.
9. von Itzstein, M.; Kiefel, M. J. In *Carbohydrates in Drug Design*; Witczak, Z. J.; Nieforth, K. A., Eds.; MDI: New York, 1997, Vol. 2, p 39.
10. Wade, R. C. *Structure* **1997**, *5*, 1139.
11. Zbiral, E. In *Carbohydrates. Synthetic Methods and Applications in Medicinal Chemistry*; Ogura, H.; Hasegawa, A.; Suami, T., Eds.; VCH: Weinheim, 1992, p 304.
12. von Itzstein, M.; Barry, J. G.; Chong, A. K. J. *Curr. Opin. Thera. Patents* **1993**, *3*, 1755.
13. Whittington, A.; Bethell, R. *Exp. Opin. Ther. Patents* **1995**, *5*, 793.
14. Colman, P. M. *Protein Sci.* **1994**, *3*, 1687.
15. Colman, P. M.; Ward, C. W. *Curr. Top. Microbiol. Immunol.* **1985**, *114*, 177.
16. Palese, P.; Tobita, K.; Ueda, M.; Compans, R. W. *Virology* **1974**, *61*, 397.
17. Liu, C. G.; Eichelberger, M. C.; Compans, R. W.; Air, G. M. *J. Virol.* **1995**, *69*, 1099.
18. Colman, P. M. In *Peptide and Protein Reviews*; Hearn, M. T. W., Ed.; Marcel Dekker: New York, 1984, p 215.
19. Varghese, J. N.; Laver, W. G.; Colman, P. M. *Nature* **1983**, *303*, 35.
20. Varghese, J. N.; Colman, P. M. *J. Mol. Biol.* **1991**, *221*, 473.
21. Baker, A. T.; Varghese, J. N.; Laver, W. G.; Air, G. M.; Colman, P. M. *Proteins: Struct. Funct. Genet.* **1987**, *1*, 111.
22. Tulip, W. R.; Varghese, J. N.; Baker, A. T.; van Donkelaar, A.; Laver, W. G.; Webster, R. G.; Colman, P. M. *J. Mol. Biol.* **1991**, *221*, 487.
23. Burmeister, W. P.; Ruigrok, R. W. H.; Cusack, S. *EMBO J.* **1992**, *11*, 49.
24. Colman, P. M.; Varghese, J. N.; Laver, W. G. *Nature* **1983**, *303*, 41.
25. Colman, P. M.; Hoyne, P. A.; Lawrence, M. C. *J. Virol.* **1993**, *67*, 2972.
26. Varghese, J. N.; McKimm-Breschkin, J. L.; Caldwell, J. B.; Kortt, A. A.; Colman, P. M. *Proteins: Struct. Funct. Genet.* **1992**, *14*, 327.
27. Bossart-Whitaker, P.; Carson, M.; Babu, Y. S.; Smith, C. D.; Laver, W. G.; Air, G. M. *J. Mol. Biol.* **1993**, *232*, 1069.
28. Luo, M.; Jedrzejas, M. J.; Singh, S.; White, C. L.; Brouillette, W. J.; Air, G. M.; Laver, W. G. *Acta Cryst. D-Biol. Cryst.* **1995**, *51*, 504.

29. Singh, S.; Jedrzejas, M. J.; Air, G. M.; Luo, M.; Laver, W. G.; Brouillette, W. J. *J Med. Chem.* **1995**, *38*, 3217.
30. Varghese, J. N.; Epa, V. C.; Colman, P. M. *Protein Sci.* **1995**, *4*, 1081.
31. Kim, C. U.; Lew, W.; Williams, M. A.; Liu, H. T.; Zhang, L. J.; Swaminathan, S.; Bischofberger, N.; Chen, M. S.; Mendel, D. B.; Tai, C. Y.; Laver, W. G.; Stevens, R. C. *J. Am. Chem. Soc.* **1997**, *119*, 681.
32. Janakiraman, M. N.; White, C. L.; Laver, W. G.; Air, G. M.; Luo, M. *Biochemistry* **1994**, *33*, 8172.
33. Lentz, M. R.; Webster, R. G.; Air, G. M. *Biochemistry* **1987**, *26*, 5351.
34. Chong, A. K. J.; Pegg, M. S.; Taylor, N. R.; von Itzstein, M. *Eur. J. Biochem.* **1992**, *207*, 335.
35. Taylor, N. R.; von Itzstein, M. *J. Med. Chem.* **1994**, *37*, 616.
36. Tiralongo, J.; Pegg, M. S.; von Itzstein, M. *FEBS Lett.* **1995**, *372*, 148.
37. Corfield, A. P.; Schauer, R. In *Sialic Acids - Chemistry, Metabolism and Function*; Schauer, R., Ed.; Springer-Verlag: Wien, 1982, p 195.
38. Khorlin, A. Y.; Privalova, I. M.; Zakstelskaya, L. Y.; Molibog, E. V.; Evstigneeva, N. A. *FEBS Lett.* **1970**, *8*, 17.
39. Suzuki, Y.; Sato, K.; Kiso, M.; Hasegawa, A. *Glycoconjugate J.* **1990**, *7*, 349.
40. Kiefel, M. J.; Beisner, B.; Bennett, S.; Holmes, I. D.; von Itzstein, M. *J. Med. Chem.* **1996**, *39*, 1314.
41. Hasegawa, A.; Morita, M.; Ito, Y.; Ishida, H.; Kiso, M. *J. Carbohydr. Chem.* **1990**, *9*, 369.
42. Meindl, P.; Tuppy, H. *Hoppe-Seyler's Z. Physiol. Chem.* **1969**, *350*, 1088.
43. Meindl, P.; Bodo, G.; Palese, P.; Schulman, J.; Tuppy, H. *Virology* **1974**, *58*, 457.
44. Palese, P.; Schulman, J. L. In *Chemoprophylaxis and Virus Infection of the Upper Respiratory Tract*; S, O. J., Ed.; CRC: Cleveland, 1977, p 189.
45. Nöhle, U.; Beau, J.-M.; Schauer, R. *Eur. J. Biochem.* **1982**, *126*, 543.
46. von Itzstein, M.; Wu, W.-Y.; Kok, G. B.; Pegg, M. S.; Dyason, J. C.; Jin, B.; Phan, T. V.; Smythe, M. L.; White, H. F.; Oliver, S. W.; Colman, P. M.; Varghese, J. N.; Ryan, D. M.; Woods, J. M.; Bethell, R. C.; Hotham, V. J.; Cameron, J. M.; Penn, C. R. *Nature* **1993**, *363*, 418.
47. von Itzstein, M.; Dyason, J. C.; Oliver, S. W.; White, H. F.; Wu, W.-Y.; Kok, G. B.; Pegg, M. S. *J. Med. Chem.* **1996**, *39*, 388.
48. Goodford, P. J. *J. Med. Chem.* **1985**, *28*, 849.
49. Taylor, N. R.; von Itzstein, M. *J. Comp-Aided Mol. Design*, **1996**, *10*, 233.
50. Holzer, C. T.; von Itzstein, M.; Jin, B.; Pegg, M. S.; Stewart, W. P.; Wu, W. Y. *Glycoconjugate J.* **1993**, *10*, 40.
51. von Itzstein, M.; Wu, W.-Y.; Jin, B. *Carbohydr. Res.* **1994**, *259*, 301.
52. von Itzstein, M.; Wu, W.-Y.; Phan, T. V.; Danylec, B.; Jin, B.; (Biota Holdings Pty Ltd) WO 91 16,320, **1991** (CA 117:49151y).
53. Pegg, M. S.; von Itzstein, M. *Biochem. Mol. Biol. Int.* **1994**, *32*, 851.
54. Crennell, S.; Garman, E.; Laver, G.; Vimr, E.; Taylor, G. *Structure* **1994**, *2*, 535.

55. Crennell, S. J.; Garman, E. F.; Laver, W. G.; Vimr, E. R.; Taylor, G. L. *Proc. Natl. Acad. Sci. USA* **1993**, *90*, 9852.
56. Woods, J. M.; Bethell, R. C.; Coates, J. A. V.; Healy, N.; Hiscox, S. A.; Pearson, B. A.; Ryan, D. M.; Ticehurst, J.; Tilling, J.; Walcott, S. M.; Penn, C. R. *Antimicrob. Ag. Chemother.* **1993**, *37*, 1473.
57. Ryan, D. M.; Ticehurst, J.; Dempsey, M. H.; Penn, C. R. *Antimicrob. Ag. Chemother.* **1994**, *38*, 2270.
58. Ryan, D. M.; Ticehurst, J.; Dempsey, M. H. *Antimicrob. Ag. Chemother.* **1995**, *39*, 2583.
59. Hayden, F. G.; Treanor, J. J.; Betts, R. F.; Lobo, M.; Esinhart, J. D.; Hussey, E. K. *J. Am. Med. Assoc.* **1996**, *275*, 295.
60. Jedrzejas, M. J.; Singh, S.; Brouillette, W. J.; Laver, W. G.; Air, G. M.; Luo, M. *Biochemistry* **1995**, *34*, 3144.
61. Chand, P.; Babu, Y. S.; Bantia, S.; Chu, N.; Cole, L. B.; Kotian, P. L.; Laver, W. G.; Montgomery, J. A.; Pathak, V. P.; Petty, S. L.; Shrout, D. P.; Walsh, D. A.; Walsh, G. M. *J. Med. Chem.* **1997**, *40*, 4030.
62. Kim, C. U.; Lew, W.; Williams, M.; Zhang, L.; Swaminathan, S.; Bischofberger, N.; Chen, M. S.; Mendel, D.; Li, W.; Tai, L.; Escarpe, P.; Cundy, K. C.; Eisenberg, E. J.; Lacy, S.; Sidwell, R. W.; Stevens, R. C.; Laver, W. G. *New Potent, Orally Active Neuraminidase Inhibitors as Anti-Influenza Agents: In Vitro and In Vivo Activity of GS 4071 and Analogues*; 36th ICAAC Meeting: New Orleans, 1996, p 171 (H44).
63. Staschke, K. A.; Colacino, J. M.; Baxter, A. J.; Air, G. M.; Bansal, A.; Hornback, W. J.; Munroe, J. E.; Laver, W. G. *Virology* **1995**, *214*, 642.
64. Blick, T. J.; Tiong, T.; Sahasrabudhe, A.; Varghese, J. N.; Colman, P. M.; Hart, G. J.; Bethell, R. C.; McKimm-Breschkin, J. L. *Virology* **1995**, *214*, 475.
65. Gubareva, L. V.; Bethell, R.; Hart, G. J.; Murti, K. G.; Penn, C. R.; Webster, R. G. *J. Virol.* **1996**, *70*, 1818.
66. McKimm-Breschkin, J. L.; Blick, T. J.; Sahasrabudhe, A.; Tiong, T.; Marshall, D.; Hart, G. J.; Bethell, R. C.; Penn, C. R. *Antimicrob. Ag. Chemother.* **1996**, *40*, 40.
67. Hayden, F. G.; Osterhaus, A. D. M. E.; Treanor, J. J.; Fleming, D. M.; Aoki, F. Y.; Nicholson, K. G.; Bohnen, A. M.; Hirst, H. M.; Keene, O.; Wightman, K. *New Engl. J. Med.* **1997**, *337*, 874.
68. Billich, A. *IDrugs*, **1998**, *1*, 122.

BICYCLAM DERIVATIVES AS HIV INHIBITORS

Gary J. Bridger and Renato T. Skerlj

Advances in Antiviral Drug Design
Volume 3, pages 161–229.
Copyright © 1999 by JAI Press Inc.
All rights of reproduction in any form reserved.
ISBN: 0-7623-0201-1

I. INTRODUCTION

Serendipity has always played a role in drug discovery. In 1989, during routine anti-HIV screening of a library of polyamine ligands and their corresponding metal complexes, it was discovered that a 2,2′-bicyclam dimer,[1] formed as a 1% impurity during a nickel-template synthesis of cyclam (1,4,8,11-tetraazacyclotetradecane) was responsible for the modest antiviral activity exhibited by a commercial sample of cyclam. Subsequently, an authentic sample of the impurity d,l-2,2′-bicyclam (AMD1657, Figure 1) was obtained and this compound exhibited potent and selective inhibition of HIV-1 and HIV-2 replication at submicromolar concentrations.[2] A medicinal chemistry program was initiated to improve the antiviral activity of AMD1657 and this initially led to the identification of AMD2763 (Figure 1), a bis-macrocycle in which the cyclam rings are connected at nitrogen (rather than carbon) positions through an n-propyl linker.[2] This analog displayed comparable antiviral inhibitory potency to AMD1657 but had the added advantage that N-linked aza-macrocycles are synthetically more accessible. However, optimum antiviral activity was exhibited by bis-cyclams connected by a phenylenebis(methylene) linker. The prototype compound, AMD3100 (also named JM3100 or SID791) (Figure 1), has an IC_{50} against HIV-1 and HIV-2 replication

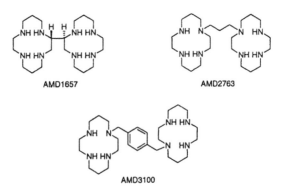

Figure 1.

of 1–10 ng/mL which is at least 100,000-fold lower than the cytotoxic concentration.[3]

We have previously reported that bicyclams inhibit HIV replication by binding to the chemokine receptor CXCR4, the coreceptor used by T-tropic (T-cell tropic) (X4) HIV viruses for membrane fusion and entry of the virus into cells of the immune system.[4–8] In fact, the first coreceptor identified to mediate HIV entry into cells was fusin,[9] later renamed CXCR4 to indicate the conserved sequence of cysteine residues in the natural ligand, the CXC-chemokine, stromal cell-derived factor 1 (SDF-1α).[10,11] CXCR4 was shown to mediate entry of T-tropic viruses and this entry was inhibited by SDF-1α. Subsequently, the CC-chemokine receptor, CCR5, was found to mediate entry of M-tropic (macrophage tropic) (R5) HIV viruses and this process was also inhibited by its natural CC-chemokine ligands, namely regulated on activation normal T-cell expressed and secreted (RANTES), and macrophage inflammatory proteins (MIP-1α and MIP-1β).[5,12–15] Despite the number of chemokine receptors which have been reported to mediate HIV entry into cells, CXCR4 and CCR5 appear to be the only physiologically relevant coreceptors used by a wide variety of primary clinical HIV strains.[16–18] These viruses use CD4 as the primary receptor on T-cells and can exclusively use CCR5, CXCR4, or both for viral entry into the cell.[19,20] Whereas CCR5 is expressed on the surface of leukocytes, particularly monocytes, and T-lymphocytes upon activation, CXCR4 is widely expressed.[21,22] For example, in addition to monocytes/macrophages,[18,23] T-lymphocytes, and neutrophils,[21] CXCR4 is expressed in human endothelial cells[24,25] and microglia, astrocytes and neuronal cells in the brain.[26–30] The coreceptor used by the virus for entry (or viral tropism) appears to correlate with disease progression. Whereas the most commonly transmitted strains are the M-tropic, non-syncytium-inducing (NSI) phenotype which utilize CCR5 for entry, the T-tropic, syncytium-inducing (SI) strains that use CXCR4 are rarely transmitted.[31–33] The X4 strains are considerably more pathogenic and their appearance during an extended period of infection correlates with a decline in CD4+ T-cell counts and a more rapid disease progression.[19,32,34,35] Primary isolates have also been classified with respect to replication rate and viremia. Non-syncytium-inducing viruses are slow/low and use CCR5, whereas the rapid/high, syncytium-inducing viruses use predominantly CXCR4.[9,14,33,36–38] Ostrowski et al. have recently shown that disease progression correlates with coreceptor expression, lower expression of CXCR4, and higher expression of CCR5 on CD4+ T-cells was indicative of advancing disease.[39] It is unclear at the present time why infection with HIV of the R5 phenotype does not lead to substantial CD4+ T-cell depletion. One recent report suggests that R5 viruses are equally cytopathic for T-cells as X4 viruses but only for CCR5+/CD4+ cells which constitute a minor portion of the total T-cell population.[40]

There has been much debate over the appropriate choice of chemokine receptors to target for the development of anti-HIV agents. Both CXCR4 and CCR5 receptors are seven trans-membrane (7TM), G-protein-coupled receptors, part of a family of receptors for the 40 or more human chemokines that have been described since the discovery of interleukin (IL)-8. Their primary role in the body is to receive regulatory signals from a specific chemokine produced locally in tissue which attract blood leukocytes and direct the maturation, trafficking, and homing of T-lymphocytes to sites of infection and inflammation.[21–22] Of the two coreceptors under study, CCR5 has received the most attention as a target for the development of antiviral agents. The rationale for this approach is threefold. First, the vast majority of HIV strains which are transmitted during infection are M-tropic and use CCR5 as a coreceptor to mediate infection.[20,31–33,35,36] Second, a certain segment of the population are either homozygous or heterozygous for a 32-base pair deletion in CCR5 which prevents functional expression of the CCR5 receptor on the cell surface. The homozygous population are resistant to initial infection upon repeated exposure to HIV, whereas the heterozygous population generally exhibits slower disease progression following seroconversion.[41–48] Despite these genetic abnormalities, patients exhibiting the CCR5 Δ32 allele are perfectly healthy. However, the heterozygous CCR5 Δ32 population does appear to be at higher risk from infection with virus of the syncytium-inducing, X4 phenotype which may explain the faster disease progression in a sub-group of these patients compared to patients that are homozygous for the CCR5 Δ32 allele.[16] Third, it has been suggested that CD4 and CCR5 are the primary receptors for tissue-infected brain and colon-derived viruses, independent of the coreceptor status (R5 or X4) of the blood-derived viruses; this may allow the establishment of viral "reservoirs" that remain a significant therapeutic barrier to complete viral eradication.[49-51]

A similar evaluation of CXCR4 as a target for the development of antiviral agents is substantially more complex. Several groups have documented the importance of the signaling mechanism provided by SDF-1α upon expression and binding to CXCR4. For example, CXCR4 or SDF-1α knock-out mice exhibit cerebellar, cardiac, and gastrointestinal tract abnormalities and ultimately die *in utero*.[52-54] CXCR4-deficient mice also display hematopoietic defects;[54] the migration of CXCR4-expressing leukocytes and hematopoietic progenitors to SDF-1α appears to be important for maintaining B-cell lineage and localization of CD34+ progenitor cells in bone marrow[55-59] (which may be mechanistically related to the early loss of hematopoietic progenitors observed in HIV-1-infected subjects).[60-62] Tumor vascularization in lung cancer is regulated by the CXC family of chemokines which can act as angiogenic or angiostatic factors depending upon the presence of the ELR (Glu-Leu-Arg) motif in their N-terminus.[63-65] SDF-1α is ELR-negative and therefore inhibits tumor growth and metastasis. These combined observations clearly indicate a very specific role for the CXCR4 receptor and its sole chemokine ligand SDF-1α in direct contrast to CCR5 which binds several ligands, obviously with some redundancy.[21]

Nevertheless, there are compelling reasons to develop antiviral agents targeted at CXCR4 in order to inhibit HIV replication. As previously mentioned, progression to X4 using syncytium-inducing HIV-1 strains is associated with a more rapid disease course and faster CD4+ T-cell decline. Inhibition of CXCR4 could select for the less pathogenic R5 viruses and delay progression to AIDS. The reciprocal approach, that is, inhibition of CCR5 and R5 virus replication, could have devastating therapeutic consequences by selecting for syncytium-inducing strains.[66] We have recently shown that combinations of X4 and R5 viruses prepared deliberately *in vitro* can be inhibited by AMD3100 to selectively yield R5 virus.[67] There are also recent reports of neuronal cell apoptosis induced by direct interaction of virus-associated gp120 with CD4 and CXCR4 (suggested as a possible cause of AIDS-related dementia) and of CXCR4-mediated CD8+ T-cell apoptosis.[28,68,69] Thus, CXCR4 antagonists may exhibit potent antiviral activity and concomitant protection of the immune system *in vivo*.

In August 1998, AnorMED Inc. initiated a Phase I clinical trial of AMD3100. Due to its poor oral bioavailability, AMD3100 was administered intravenously or subcutaneously to 12 volunteers in order to obtain initial safety and pharmacokinetic data in humans. The drug was well-tolerated and exhibited linear pharmacokinetics. However, the identification and clinical development of a CXCR4 or CCR5 antagonist that is orally absorbed may be advantageous to any drug that targets a chemokine receptor for three reasons: (1) the currently approved drugs that target, for example, reverse transcriptase (AZT, DDI) or HIV protease (Crixivan, Ritonavir, Saquinavir) are all administered orally[70]; (2) the currently approved drugs are administered in combination, which not only provides superior antiviral efficacy but also reduces the rate of resistance development to individual drugs; (3) independent of the target selected (CXCR4 or CCR5), there is no obvious reason to believe that a chemokine antagonist could not be given in combination. In view of the fact that a selective chemokine antagonist would inhibit a specific viral population based on phenotype,[66,67] the coadministration with currently approved antiviral agent(s) that are nonphenotype-selective may even be compulsory.

The medicinal chemistry group at AnorMED is pursuing a second generation analog of AMD3100 that is efficiently orally absorbed. This is a formidable task since the azamacrocyclic rings of AMD3100 contain all of the structural features which are an impediment to good oral absorption. The drug is highly charged at physiological pH (the overall charge is +4), it contains large numbers of hydrogen-bond donors and acceptors, and has a large surface area and high water solubility (low partition coefficient). As a general approach to solving this problem, we initially focused on defining the structural features of AMD3100 responsible for potent anti-HIV activity and then attempted to replace the secondary amines groups with neutral heteroatoms or amines of lower basicity. This chapter will profile AMD3100, and describe our early efforts to identify compounds with improved oral bioavailability.

II. MECHANISM OF ACTION

It was determined early in the discovery process that bicyclam antiviral agents interfered in the early steps of the HIV-viral replication process, postbinding of virus-derived gp120 to CD4 on the surface of T-cells but before reverse transcription can occur.[2,3,71] This novel mechanism of action was deduced from several experiments in which bicyclams AMD3100 (and AMD2763) were compared to other known anti-HIV agents. For example, AMD3100 failed to block HIV-1 binding to cells and to inhibit binding of an anti-gp120 monoclonal antibody to persistently infected HUT-78 cells under conditions that dextran sulfate effectively inhibited binding. AMD3100 was also not inhibitory to HIV reverse transcriptase or HIV protease, the molecular targets for the currently approved chemotherapeutic agents. In order to more accurately determine the "stage" at which bicyclams were intervening in the HIV replicative cycle, a time-of-addition experiment was developed.[3] In this assay, cells were infected at high virus multiplicity to ensure that virus replication was synchronized in the whole cell population and test compounds were then added at varying times (in hourly intervals) to the infected cells. The relative stage of mechanistic intervention of the bicyclams was then determined by the time at which antiviral efficacy was lost, since addition of the bicyclams after the mechanistic target has been utilized for virus replication renders the compound ineffective and p24 antigen production increases dramatically, as shown in Figure 2.

In a similar manner, AMD2763 and AMD3100 were determined to interact at a stage following the virus adsorption step (at which dextran sulfate interacts) but preceding reverse transcription and subsequent processing by HIV protease [compare to AZT; TIBO R82913 (a non-nucleoside RT inhibitor); and Ro 31-8959 (Saquinavir, an HIV protease inhibitor), respectively]. In addition, HIV-1 strain NL4.3 made partially resistant to AMD3100 exhibited several amino acid mutations in gp120.[72,73] Thus, AMD2763 and AMD3100 were assumed to interact with an HIV uncoating/fusion process and the molecular target for bicyclam intervention at the HIV inhibitory step seemed most likely to be virus-associated gp120. To further support this tentative mechanism, AMD3100 was tested for its ability to inhibit virus-induced syncytium formation between cocultures of persistently HIV-1 (III_B)-infected HUT-78 cells and uninfected MOLT-4 cells. It was reasoned that syncytium formation between HIV-infected and uninfected cells could be viewed as an assay for virus–cell fusion. Although AMD3100 indeed inhibited syncytium formation, the effec-

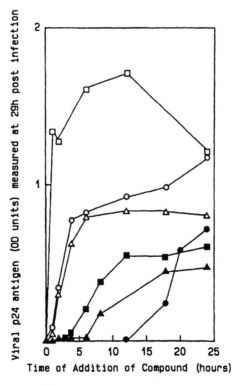

Figure 2. Time-of-addition experiment.[3] MT-4 cells were infected with HIV-1(III$_B$) at an MOI of >1, and the test compounds were added at different times postinfection. Viral p24 antigen production was determined at 29 h postinfection and is expressed as OD (optical density) units. Symbols: □, dextran sulfate (molecular weight, 5000) (10 μM); ○, AMD2763 (25 μM); △, AMD3100 (0.15 μM); ■, AZT (0.2 μM); ▲, TIBO R82913 (0.5 μM); ●, Ro 31-8959 (6.5 nM).

tive concentration was significantly higher (>260-fold) than the concentration of AMD3100 required to inhibit HIV-1(III$_B$) replication.[3]

However, the precise molecular target for bicyclam intervention in the HIV-replicative cycle was not elucidated until 1997, following the discovery that HIV required a coreceptor, in addition to CD4, for viral entry into cells. In view of the burgeoning literature on the role of chemokine receptors in HIV infection and the early stage at which they (and bicyclams) are involved in the HIV replication process, the ability of AMD3100 to interfere with virus entry into cells via the chemokine receptors CXCR4 and CCR5 was evaluated. Studies on the direct interaction of AMD3100 with the coreceptors clearly demonstrated that the bicyclam bound to CXCR4 with high affinity, and at concentrations that were comparable to the effective concentration of AMD3100 required to inhibit HIV-1 and HIV-2 replication.[4–6] However, AMD3100 did not interfere with CCR5.[4–7,18] Serendipitously, therefore, we had rediscovered AMD3100, a potent and selective antagonist of CXCR4 and the first small-molecule inhibitor of HIV-1 entry via inhibition of a chemokine receptor. In mechanistic experiments, AMD3100 displayed all the characteristics of an antagonist: AMD3100 dose-dependently inhibited binding of 12G5 (a CXCR4-specific monoclonal antibody) to CXCR4 on lymphocytic SUP-T1 cells, completely inhibited binding of ^{125}I-labeled SDF-1α to the MT-2 cell line, and dose-dependently inhibited the signal transduction (indicated by an increase in calcium flux) in response to SDF-1α in SUP-T1 (see Figure 3) and monocytic THP-1 cells.[5] In addition, AMD3100 completely blocked the response of THP-1 cells to the chemotactic effects of SDF-1α. However, unlike SDF-1α, AMD3100 alone did not directly induce a calcium flux or downregulate the receptor. We have recently reported a structure–activity relationship study on a diverse series of bicyclam analogs which shows a close correlation between antiviral potency against X4 strains of HIV-1, inhibition of 12G5 binding to CXCR4, and inhibition of calcium flux induced by SDF-1α.[8]

The specificity of chemokine receptor inhibition by AMD3100 was confirmed by Alizon et al.[7] using a panel of cell lines transfected with a family of chemokine receptors reported to mediate infection of HIV and SIV strains. While AMD3100 dose-dependently inhibited HIV-1 (LAI, NDK, 89.6) and HIV-2 (ROD) in U373MG-CD4 cells stably transfected with the chemokine receptor CXCR4, AMD3100 failed to inhibit infection of cells stably transfected with the chemokine receptors CCR5, CCR3, BOB, and Bonzo, and cells transiently expressing US28. To further characterize the interaction of AMD3100 with CXCR4 at the molecular level, a series of mutant CXCR4 receptors were generated containing single amino acid mutations or domain deletions. Single substitutions of neutral amino acids for aspartic acid (D182G and D193A) in the second extracellular loop (ECL2) or

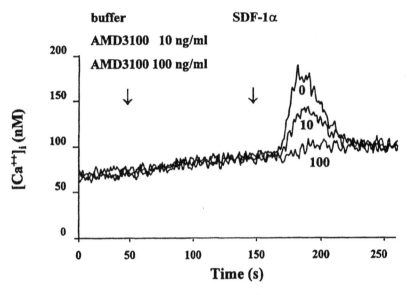

Figure 3. Inhibition of SDF-1α induced Ca^{2+} flux in SUP-T1 cells by AMD3100 pretreatment at 100 ng/mL and 10 ng/mL.[5] SDF-1α was given as a second stimulus (right arrow) at 30 ng/mL.

nonaromatic amino acids for phenylalanine (F172A) in the adjacent fourth trans-membrane spanning domain (TM4) of CXCR4 gave receptors that support HIV-1 entry but were relatively resistant to inhibition by AMD3100 (Figure 4). Consequently, the aspartic acid residues of ECL2 were proposed to interact with the cyclam rings of AMD3100 during binding and inhibition through electrostatic inter-actions.[7] Several other residues were found to be important for AMD3100 binding and inhibition of HIV-1 entry, including D171 and F174 in TM4 and F199 and F201 in ECL2. These amino acids are components of, or proximal to, the Phe-X-Phe motifs of CXCR4, which were suggested as possible binding site(s) for the aromatic linker of AMD3100.[7] However, a mutant CXCR4 receptor in which the majority of the amino-terminal extracellular domain had been deleted (amino acids 4–36) remained sensitive to inhibition of HIV-1 infection by AMD3100.[7]

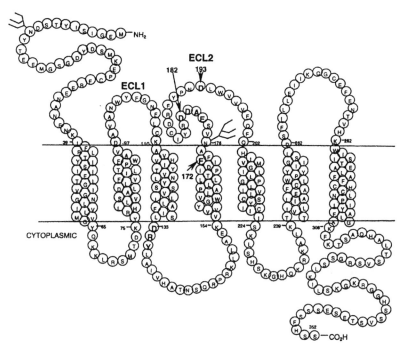

Figure 4. Sequence of the HIV-1 coreceptor, CXCR4. Reproduced with permission from Doranz et al. *J. Virol.* **1999**, *73*, 2752–2761. Copyright 1999 by the American Society for Microbiology.

The overwhelming experimental evidence now clearly supports our proposition that the mechanism by which AMD3100 inhibits HIV-1 and HIV-2 replication is by antagonism of the chemokine receptor CXCR4. Although the precise binding site for AMD3100 on CXCR4 during the HIV-inhibitory step remains elusive, a recent study further implicates the ECL2 domain of CXCR4 in binding of, and signaling by, SDF-1α.[75] For example, replacement of the highly negatively charged amino acid sequence Glu-Ala-Asp-Asp (EADD) in ECL2 with Gln-Ala-Ala-Asn (QAAN) gave a mutant CXCR4 receptor which binds SDF-1 with high affinity but failed to Ca^{2+} signal and did not function as a coreceptor for viral entry. As mentioned above, mutation of the single aspartic acid residue at position 182 (EAD**D**) in ECL2 confers relative resistance to inhibition by AMD3100 and abrogated binding of the CXCR4-specific monoclonal antibody

12G5, which recognizes an epitope containing ECL1 and ECL2 that is conformation-dependent.[7]

Other important signaling sequences of CXCR4 include the amino terminus, the third ECL, and the cytoplasmic second intracellular loop Asp-Arg-Tyr motif (the DRY box), which is highly conserved among other G-protein-coupled receptors including CCR5. Mutation of this region to Asn-Ala-Ala in CXCR4 significantly reduces the ability of the mutant CXCR4 to signal in the presence of SDF-1α. These observations are in agreement with the model of interaction between SDF-1α and CXCR4 (independently proposed by Crump et al.[76] and Heveker et al.[77]) in which amino acids 12–17 of SDF-1α (the RFFESH motif) mediate binding to the amino terminus of CXCR4 and the amino acids Lys-Pro at the N-terminus of SDF-1α activate CXCR4 by interaction with ECL2.

Interestingly, the ECL2 of feline-CXCR4, in addition to contributions from the first and third loops, appear to be important for infection by feline immunodeficiency virus (FIV).[78] Mutation of the DRY box in the second intracellular loop of feline CXCR4 to NAA gives a receptor that is still able to support infection by FIV. Clearly, the ability of feline CXCR4 to function as a coreceptor for FIV is distinguishable from its ability to G-protein signal. In view of the sequence homology between human and feline CXCR4, however, a series of bicyclams, including AMD3100, were tested for their ability to inhibit FIV replication in Crandell feline kidney (CRFK) cells. Both AMD3100 and human SDF-1α were found to inhibit FIV and HIV replication at comparable 50% effective concentrations. Furthermore, AMD3100 inhibited binding of a monoclonal antibody to the ECL2 of human and feline CXCR4 in a dose-dependent manner.[79]

Given the definitive mechanism of action, it is always easier to retrospectively interpret data that led to the assumption that virus-associated gp120 was the molecular target for bicyclam intervention at the HIV inhibitory step. As previously described, the primary evidence in support of gp120 was the accumulation of amino acid mutations in a strain of HIV-1 NL4.3 made resistant to AMD3100. In fact, marked resistance to AMD3100 was only observed after prolonged passaging of the virus in the presence of increasing concentrations of AMD3100. DNA sequencing of the AMD3100-resistant NL4.3 strain indicated that up to 11 amino acid mutations may be required to achieve resistance to bicyclams. We have subsequently found that HIV-1 NL4.3 AMD3100-resistant strain is also resistant to inhibition by SDF-1, and conversely, SDF-1 resistant HIV-1 NL4.3 is 10-fold less sensitive to inhibition with AMD3100 and does not show a switch (X4 to R5) in coreceptor use.[80] Since we now know the precise molecular target for bicyclam intervention, that is, via inhibition of the chemokine receptor CXCR4, the difficulty in generating AMD3100-resistant virus would now seem somewhat logical. Although there are reports of laboratory-adapted HIV-1 and HIV-2 strains that can infect target cells via a CD4-independent mechanism,[81] the well-defined first step of the infection process for the vast majority of primary HIV strains is binding of the HIV-associated envelope glycoprotein gp120 to CD4 on the surface of the target cell.[20] A conformational change then occurs allowing interaction of the V3 loop of gp120 with the respective chemokine receptor(s) and sub-

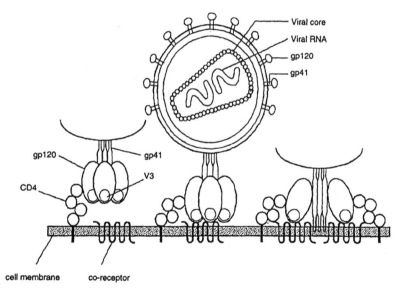

Figure 5. Model of HIV-1 entry into CD4+ cells. Following binding of the HIV-1 *env* glycoprotein gp120 to CD4, a conformational change occurs allowing interaction of the V3 loop of gp120 with the coreceptor CXCR4 and/or CCR5. Fusion and entry of the virus into the cell is also mediated by virus-associated gp41.

sequent promotion of fusion and entry of the virus into the cell mediated by the HIV-1 trans-membrane glycoprotein, gp41[20,82] (Figure 5). Recently, a synthetic peptide inhibitor (T20), corresponding to a 36-amino acid fragment of gp41, was shown to potently suppress HIV-1 replication in HIV-infected subjects.[83]

In support of the proximate feasibility of a gp120/CD4/chemokine receptor complex, Lapham et al. reported a constitutive association between CD4 and the respective coreceptors in monocytes and macrophages based on the observation that CD4 coprecipitates with CCR5 and CXCR4 monomers.[23] Direct, CD4-dependent binding of X4 gp120s to CXCR4 has also been demonstrated to occur via a ternary complex with CXCR4.[75] If one assumes that AMD3100 is effectively competing with a protein sequence of the viral envelope during the formation of a ternary complex with CXCR4, then the selective pressure exhibited by AMD3100 on the virus can be considered secondary (cellular) rather than primary (viral) and may be responsible for the slow rate of resistance development to bicyclams.

III. ANTIVIRAL ACTIVITY PROFILE OF AMD3100

AMD3100 is active against a variety of laboratory and clinical T-tropic strains of HIV-1 such as III$_B$, RF and NL4.3, and HIV-2 (ROD) with 50% inhibitory concentrations (IC$_{50}$'s) in the range of 2–7 ng/mL [5] (Table 1). Consistent with its mechanism of action, AMD3100 was inactive against M-tropic HIV viral strains including BaL and SF-162 or simian immunodeficiency virus (SIV) strains such as MAC-251 in MT-4 cells and AGM-3 in MOLT-4 cells. These strains of HIV and SIV (in human cells) use exclusively CCR5 as a coreceptor for viral entry. In identical assays, the chemokine SDF-1 inhibited the replication of T-tropic HIV-1 strains with IC$_{50}$'s of 20–100 ng/mL but was inactive against M-tropic strains. As expected, the CCR5 ligand RANTES gave the opposite activity profile for inhibition of T- and M-tropic viruses. AMD3100 was also active against HIV strains resistant to 3'-azido-3'-deoxythymidine (AZT) and 2',3'-dideoxyinosine (DDI) at similar concentrations to those exhibited against the wild-type virus.[84]

Again consistent with its mechanism of action, AMD3100 was inhibitory to immunodeficiency virus strains that use CXCR4 for viral entry. For example, the simian immunodeficiency strain SIVmnd(GB-

Table 1. The Anti-HIV Activity Profile of AMD3100 Correlated with Coreceptor Use

		IC_{50} (ng/mL)		
Strain	*Coreceptor used*	*AMD3100*	*SDF-1α*	*RANTES*
T-tropic				
HIV-1 IIIB	CXCR4	2	20	>1,000
HIV-1 RF	CXCR4	5	50	>1,000
HIV-1 NL4-3	CXCR4	3	100	>1,000
HIV-2 ROD	CXCR4	7	55	>1,000
M-tropic				
HIV-1 BaL	CCR5	>25,000	>1,000	25
HIV-1 SF-162	CCR5	>25,000	>1,000	5
HIV-1 ADA	CCR5 (CCR2b, CCR3)	>25,000	>1,000	10
HIV-1 JR-FL	CCR5 (CCR2b, CCR3)	>25,000	>1,000	4

1), isolated from the mandrill, replicates in a human T-cell (CEM) line and only in a transfected cell (HOS.CD4) line expressing CXCR4. This SIV strain was inhibited by human SDF-1α and AMD3100 (IC_{50} = 8 ng/mL).[85] As previously mentioned, bicyclams are also potent inhibitors of feline immunodeficiency virus (FIV) that has been adapted to replicate in Crandell feline kidney (CRFK) cells.[79,86] The IC_{50}'s for a series of bicyclams, including AMD3100, against FIV replication in CRFK cells were comparable to their IC_{50}'s against HIV-1 III_B replication in the human CD4+ MT-4 cell line and both FIV and HIV replication was inhibited by SDF-1α. Furthermore, six primary FIV isolates were equally susceptible to inhibition by AMD3100 in feline thymocytes. The comparable structure–activity relationship displayed by bicyclams against HIV and FIV suggests that CXCR4 is an essential coreceptor for FIV infection even though CD4 is not the primary receptor.

Finally, in order to demonstrate that inhibition of viral entry into cells via antagonism of CXCR4 is a legitimate target for antiviral chemotherapy of HIV infection, the inhibitory effect of AMD3100 was evaluated in intrathymically HIV-1-infected SCID-hu (Thy/Liv) mice.[87] In this model, once daily, subcutaneous injections of AMD3100 at nontoxic doses caused a reduction in p24 antigen expression, a dose-dependent decrease in viremia and a protection of the

decrease in CD4/CD8 T-cell ratio. Antiviral efficacy was also poten-
tiated by the combined administration of AMD3100 with AZT or DDI.
Thus, AMD3100 exhibits potent antiviral efficacy *in vitro* and *in vivo*,
clearly establishing inhibition of CXCR4 and virus entry into cells as
a target for chemotherapeutic intervention with antiviral agents.

IV. DRUG DESIGN AND STRUCTURE–ACTIVITY RELATIONSHIPS OF ANTI-HIV BIS-AZAMACROCYCLES

Based on our initial lead, AMD1657, we sought to determine the main
structural features responsible for antiviral activity with the ultimate
goal of increasing anti-HIV potency. To this end, a survey of the
chemical literature revealed that several bis-tetra-azamacrocycles
connected at a single nitrogen position had been exemplified and were
readily synthetically accessible (Table 2). The most prominent example was AMD2763, a bis-cyclam in which the macrocyclic rings were
connected at nitrogen through an *n*-propylene linker. This compound
inhibited HIV-1 and HIV-2 replication with 50% effective concentrations of 0.248 and 1.0 μM, respectively, which is more than 2500-fold
lower than its 50% cytotoxic concentration in MT-4 cells.[2] Using
similar chemical routes we prepared the ethyl- (AMD2762) and
hexyl-linked (AMD2849) analogs to determine the optimum distance
between the macrocyclic rings. Antiviral testing of these compounds
gave us our first clue as to the specificity of the receptor: while the
hexyl analog exhibited comparable antiviral potency to AMD1657 and
AMD2763, the ethyl analog was significantly less potent for inhibition
of HIV-1 and HIV-2 replication.[2] Given the structure of our initial lead
AMD1657 in which the macrocyclic rings are connected at carbon
positions, we also spent considerable synthetic effort preparing an
analog of AMD2763 in which the cyclam rings were connected
unsymmetrically; that is, via carbon and nitrogen positions of the rings
through an *n*-propyl linker.[88] AMD2936 was intermediate in antiviral
potency between AMD2762 and the hexyl-linked analog AMD2849
(Table 2).

However, a significant increase in activity was only achieved when
we realized that the six-carbon chain of the hexyl linker in AMD2849
was equivalent in carbon–carbon bond connectivity to a phenylene-
bis(methylene) linker. This group reduces the number of bond rota-

Table 2.

AMD No.	Structure	EC$_{50}$ (μM) HIV-1 (III$_B$)	EC$_{50}$ (μM) HIV-2 (ROD)	CC$_{50}$ (μM)
1498		399	150	1248
1657		0.144	1.01	319
2936		1.96	3.56	>445
2762	n = 2	18.6	124.6	349
2763	n = 3	0.248	1.00	>622
2849	n = 6	0.616	0.616	290

tions to four and adds an aromatic group to increase the hydrophobic character. The prototype analog AMD3100, which connects the macrocyclic rings at *para* positions was 2 orders of magnitude more potent than AMD2763 for inhibition of HIV-1 and HIV-2 replication in MT-4 cells.[3] The mechanism of action and activity profile of this compound has already been described in some detail. Although we were encouraged by the improved antiviral potency of AMD3100, the results raised some inevitable questions regarding the structure and function of the macrocyclic rings, and the group connecting them. For example, what is the correct configuration and size of the macrocyclic ring required for potent activity? What are the effects of substituents on the aromatic linker or the macrocyclic ring? What role, if any, does the metal chelating ability of the macrocyclic rings play in the antiviral activity of this compound? In order to design more potent compounds with the appropriate biopharmaceutical properties, we deemed it necessary to have definitive answers to all of these questions.

To answer the specific question of correct ring size, synthetic methodology was developed to prepare all ring sizes of between 9- and 22-ring members, containing 3–5 nitrogens per ring and connected via a phenylenebis(methylene) linker in *ortho*, *meta*, or *para* positions.[89] The anti-HIV activity of a representative group of bis-macrocyclic compounds is shown in Table 3. It was apparent that high antiviral potency is limited to compounds in which the macrocycles contain four nitrogen atoms per ring and 12–14 ring members.

One interesting structural feature we are still unable to fully explain is the combined effect of macrocyclic ring size and linker substitution. Whereas in the series of 14-membered tetra-aza macrocyclic dimers, *para* substitution of the aromatic linker is required for potent anti-HIV-1 and HIV-2 activity (compounds **5–8**, Table 3), dimers of the 12- and 13-membered rings require a *meta*-substituted linker for optimum

Table 3.

CPD#	Structure	Phenyl Substitution	EC$_{50}$ (μM) HIV-1 (III$_B$)	EC$_{50}$ (μM) HIV-2 (ROD)	CC$_{50}$ (μM)
1		para	0.3218	2.3600	>55
2		meta	0.0751	0.5364	>20
3		para	0.1668	0.2341	>216
4		meta	0.0408	0.0618	>191
5		para	0.0253	0.0590	>421
6		meta	0.3226	0.6451	>403
7		para	0.0042	0.0059	>421
8		meta	0.0337	0.0422	>421
9		ortho	1.3574	3.1279	>168

activity (compounds **1–4**, Table 3). It is difficult to envisage a structural overlap between these seemingly nonsuperimposable groups of structures. We can only assume that the CXCR4 receptor is sufficiently flexible to accommodate relatively small variations in the manner in which functional groups of the inhibitor(s) are presented. In order to resolve this issue, *meta*-linked bis-[12]aneN$_4$ analog **2** was recently tested for its ability to inhibit 12G5 binding to CXCR4 and then compared to AMD3100. It was reasoned that if both analogs were binding to the same portion of the CXCR4 receptor then the 17-fold lower EC$_{50}$ for inhibition of HIV-1 replication exhibited by AMD3100 (**7**) compared to compound **2** should be approximately reflected in a corresponding reduction in the EC$_{50}$ for 12G5 binding. However, **2** was an extremely poor inhibitor of 12G5 binding which may indicate a different mode of binding compared to AMD3100 and subsequently, the 12G5 antibody is still able to recognize its desired epitope.

The aromatic group which connects the cyclam rings in AMD3100 has been proposed to interact with Phe-X-Phe motifs of ECL2 or TM-4 of the CXCR4 receptor.[7] In this regard we have previously prepared and reported the anti-HIV activity of a series of AMD3100 analogs containing electron-withdrawing or electron-donating groups on the aromatic ring.[89] Activity appears to be insensitive to the electronic influence of the substituent but sterically demanding groups markedly reduce antiviral potency. This structure–activity relationship is exemplified by the data presented in Tables 4 and 5.

The introduction of methoxy, methyl, bromo, or fluoro groups onto the aromatic ring of AMD3100 (**7**) had little effect on the anti-HIV-1 and HIV-2 activity (compare compounds **10–14** to **7**, Table 4), whereas a bromo group at position 2 of the *meta*-linked bicyclam analog **16** significantly reduced activity (Table 5). That the reduction in activity of the 2-bromo analog **16** was not due to the electron-withdrawing capability of a substituent in this particular position was demonstrated by compounds **15** and **17**. The 5-bromo (**15**) and 2-fluoro (**17**) analogs displayed comparable activity to the parent *meta*-linked bicyclam analog **8**. In contrast to the effects of substitutions on the aromatic linker, the introduction of fluoro or methyl groups on the cyclam ring itself was detrimental to activity (Table 6).

Presumably the *gem*-dimethyl groups in compound **19** impose an unfavorable conformation on the cyclam ring and thereby disrupt

Table 4.

		EC_{50} (μM)		
CPD#	X	HIV-1 (III$_B$)	HIV-2 (ROD)	CC$_{50}$ (μM)
10	2,5-dimethyl	0.0064	0.0011	>208
11	2,5-dichloro	0.0107	0.0025	>58
12	2-bromo	0.0061	0.0035	>203
13	2,5-dimethoxy	0.0058	0.0066	>206
14	2,3,5,6-tetrafluoro	0.0079	0.0079	>47
7	H (AMD3100)	0.0042	0.0059	>421

high-affinity binding of this compound to the CXCR4 receptor. This analog was also considerably more cytotoxic to MT-4 cells. However, the reduced activity of the fluoro analog **18** is most likely electronic in nature since the fluorine and hydrogen groups have similar ionic radii but the basicity of the amines in the fluorinated cyclam ring are reduced. The significance of macrocyclic ring protonation will be explored in more detail.

The propensity of azamacrocyclic rings such as cyclam to chelate transition metal ions with high thermodynamic stability prompted us to prepare a series of metal complexes of AMD3100 and evaluate their antiviral inhibitory capabilities[8,89] (Table 7). Complexes of lower kinetic stability, such as the zinc(II) and nickel(II) complexes (compounds **20** and **21**), exhibited comparable (or better) EC_{50}'s than the uncomplexed analog, AMD3100 (**7**), whereas complexes of high kinetic stability such as the bis-palladium complex **24** were ineffective. In addition, the anti-HIV activity displayed by these compounds paralleled their ability to inhibit 12G5 binding and inhibit the calcium flux induced by SDF-1α, which indicates they are binding to the same portion of the CXCR4 receptor.

These results are intriguing and puzzling for several reasons: while there is a distinct possibility that inhibition is transition metal-medi-

Table 5.

A   B

CPD#	Structure	X	HIV-1 (III$_B$)	HIV-2 (ROD)	CC$_{50}$ (µM)
			EC$_{50}$ (µM)		
7	A	H (AMD3100)	0.0042	0.0059	> 421
8	B	H	0.0337	0.0422	> 421
12	A	2-bromo	0.0061	0.0035	> 203
15	B	5-bromo	0.0845	0.0538	> 192
16	B	2-bromo	0.1383	0.2459	> 144
17	B	2-fluoro	0.0347	0.0734	> 201

ated via a ternary complex with CXCR4 and AMD3100 *in vitro*, it is difficult to reconcile the apparent lack of specificity inferred by the equal activity seen with the free ligand and its di-Zn and di-Ni complexes. Zinc is the most likely candidate for coordination to the macrocyclic rings *in situ* since this is the most abundant, freely

Table 6.

CPD#	R^1	R^2	R^3	R^4	HIV-1 (IIIB)	HIV-2 (ROD)	CC50 (µM)
					EC$_{50}$ (µM)		
7	H	H	H	H	0.0042	0.0059	> 421
18	F	H	H	H	0.0178	0.0121	> 203
19	H	H	Me	Me	0.1956	0.2619	17.0

Table 7.

CPD# (bound metal)	EC_{50}^{a} for HIV-1 (III$_B$) (μM)	IC_{50}^{b} 12G5 binding (μM)	$IC_{50[Ca^{2+}]_i}^{c}$ (μM)
7 (AMD3100)	0.0108	0.012	0.006
20 (Zn)	0.0077	0.0009	0.0029
21 (Ni)	0.0078	0.0157	0.0019
22 (Cu)	0.0436	0.1818	0.0454
23 (Co)	0.7697	0.5200	0.6240
24 (Pd)	57.881	10.544	59.046

Notes: [a]EC_{50}: 50% effective concentration, or concentration of the compound required to inhibit HIV-1 replication by 50%, as measured by the MTT assay, in MT-4 cells.

[b]IC_{50}: 50% inhibitory concentration, or concentration of the compound required to inhibit by 50% the binding of mAb 12G5 to CXCR4-positive SUP-T1 cells.

[c]$IC_{50[Ca^{2+}]_i}$: Concentration of the compound required to inhibit by 50% the intracellular Ca^{2+} concentration induced by SDF-1α in SUP-T1 cells.

available transition metal ion in the body. Nickel, on the other hand, is not an essential element and is categorically absent in cell culture. A simple explanation to account for the potent activity of the zinc and nickel complexes of AMD3100 is that the metal ions are removed by trans-chelation with coordinating amino acid residues present in cell culture. Irrespective of the metal complex tested, this scenario would effectively liberate some "free" AMD3100 (depending upon the stability of the metal complex), and the EC_{50}'s observed would be wholly accounted for by the concentration of AMD3100 formed in solution. While this explanation is straightforward, it is inconsistent with the chemical literature. Complexes of cyclam with zinc(II), nickel(II), and copper(II) are *kinetically* labile but nevertheless exhibit stability constants indicative of high *thermodynamic* stability. For example, the stability constants (log K's) of complexes with cyclam are reported to be in the range: 15.0–15.3 (Zn); 19.9–20.3 (Ni); and 26.4–27.9 (Cu).[90] Put another way, [Cu(cyclam)]$^{2+}$ is not decomposed in 6M HCl over a period of weeks[91] and [Ni(cyclam)]$^{2+}$ dissolved in 1M HClO$_4$ has a half-life of approximately 30 years.[92] Ironically, the stability of [Ni(cyclam)]$^{2+}$ is best illustrated by the Ni-template synthesis of cyclam: during this preparation, the nickel can only be removed from the cyclam ring in the final step by boiling the complex with excess sodium cyanide.[1,93] Since it is unlikely that a ligand set capable of forming a complex with comparable or higher stability is present in the culture medium, one must conclude that the metal complexes remain intact in the *in vitro* experiments.

Based on the assumption that metal complexes of AMD3100 remain intact during binding and inhibition of CXCR4-mediated HIV entry, we sought an alternative mechanistic interpretation of the results. If AMD3100 and its corresponding zinc and nickel complexes bind to CXCR4 with equal affinity, what common structural features does the receptor see? Since AMD3100 is unable to "coordinate" like a transition metal complex, how can the cyclam ring alone interact with protein residues in a similar manner? To potentially answer these questions we needed to examine the structural characteristics in more detail.

The cyclam ring has four pK_a's measured at 11.5, 10.3, 1.6, and 0.9, respectively, which provide an overall charge on the macrocyclic ring at physiological pH of +2.[90] The third and fourth protonations become increasingly difficult due to the repulsive effects of the two positive charges already confined in a relatively constrained macrocyclic framework. An X-ray and neutron diffraction structure of a cyclam complex with 4 equiv of 4-*t*-butylbenzoic acid confirms that the protonated ring adopts a thermodynamically favorable pseudo-*trans*-III cyclam configuration (if the center of symmetry in cyclam is treated as an atom) in which diagonal amines (positions 1,8-) are protonated.[94] This structure also revealed that the protonated cyclam ring has the propensity to form a direct complex with a carboxylic acid group. As depicted in Figure 6 each of two molecules of 4-*t*-butylbenzoic acid were in direct contact with the cyclam ring on opposing faces of the macrocycle, connected through a network of three non-equivalent hydrogen bonds spanning the protonated amines. Protonated hexaazamacrocycles form complexes with dicarboxylates in aqueous solution, presumably through similar intermolecular recognition motifs.[95,96]

This work is in agreement with an earlier structure of cyclam dihydroperchlorate salt in which the protonation sequence and cyclam ring conformation were found to be identical, and in this case, the oxygen atoms of the perchlorate group were suggested to participate in hydrogen-bonding interactions.[97] Cyclam complexes of Zn(II), Ni(II), and Cu(II) also adopt the thermodynamically favored *trans*-III cyclam configuration to give a compound with an overall charge of

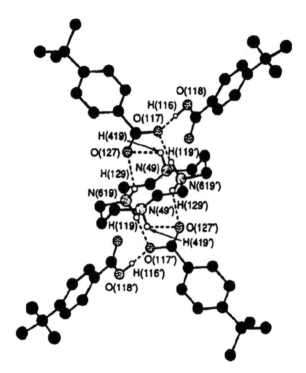

Figure 6. Crystal structure of a cyclam complex with 4 molecules of 4-*t*-butyl benzoic acid. Reprinted with permission from Adam et al. *J. Chem. Soc., Chem. Commun.* **1994**, 1539. Copyright 1994 by the Royal Society of Chemistry.

+2.[1,98–100] In striking similarity to the aforementioned doubly protonated cyclam, these metal complexes (and transition metal complexes of other tetraaza-macrocycles) can also form complexes with carboxylic acids. For example, Ito et al. has shown that a bis-zinc complex of an *ortho*-phenylenebis(methylene)-linked bicyclam derivative forms a complex with face-to-face stacking of the *trans*-III cyclam conformers in which a carbonate group (CO_3^{2-}) is coordinated between the zinc ions.[98] The crystal structure of this complex is shown in Figure 7.

The geometry around the individual zinc ions is square–pyramidal and the Zn ions are 0.425 and 0.445 Å, respectively, above the plane of the four nitrogen atoms in the macrocyclic ring. The orientation of

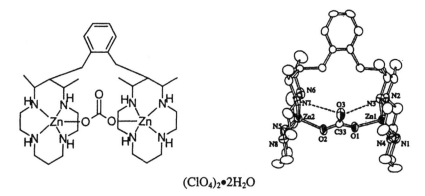

(ClO$_4$)$_2$•2H$_2$O

Figure 7. Molecular and crystal structure of an aromatic linked bicyclam derivative (bis-zinc complex). The cyclam rings are arranged face-to-face with a bridging carbonate group complexed to both zinc ions. Reprinted with permission from Kajiwara et al. *Inorg. Chem.* **1993**, *32*, 4990. Copyright 1993 by the American Chemical Society.

the CO_3^{2-} group is dominated by coordination of a unidentate C–O$^-$ to zinc and hydrogen-bonding interactions between the secondary amine protons of the cyclam ring and the noncoordinated oxygen of the carbonate group. A similar host–guest interaction of a 1,3,5-benzenetricarboxylate group with a nickel complex of a cyclam derivative has recently been reported to give a highly ordered three-dimensional network.[101]

The corresponding Zn(cyclen) complex (cyclen = 1,4,7,11-tetraazacyclododecane = [12]aneN$_4$) forms host–guest complexes with "imide"-like nucleoside analogs,[102,103] as shown in Figure 8. Although the coordinated imide was shown to be deprotonated, the secondary amine protons on the macrocyclic ring could not be definitely proven to participate in direct hydrogen-bonding interactions with the guest imide oxygen atoms due to the long N–H...O=C bond distances (2.8–3.1 Å, based on the calculated positions of the hydrogen atoms in the crystal structure). Presumably, this is due to the considerable deviation of the zinc atom above the plane of the four nitrogen atoms in the aza-macrocyclic ring of cyclen (compared to the larger cyclam ring) which places the hydrogen-bond donors (N–H groups of the azamacrocyclic ring) at a greater intermolecular distance. A variety of compounds containing single and multiple Zn(cy-

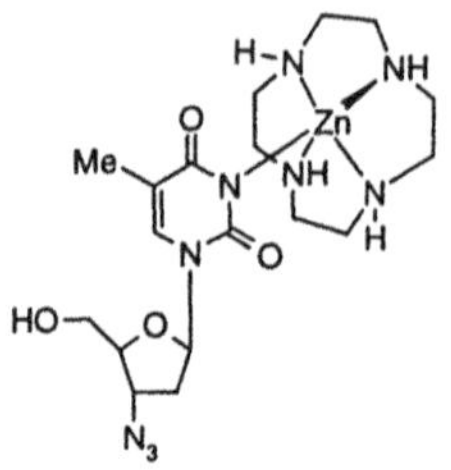

Figure 8. Complex of Zn(cyclen) with AZT. Reprinted with permission from Shionoya et al. *J. Am. Chem. Soc.* **1993**, *115*, 6730. Copyright 1993 by the American Chemical Society.

clen) macrocyclic complexes have also been shown to bind phosphate via unidentate coordination of the anionic P–O⁻ group[104,105] (Figure 9). The bis-zinc(II) complex of the *meta*-linked phenylene-bis(methylene) linked bis-cyclen analog (Figure 10) was also found to bind barbital by double deprotonation. Coincidentally, we, and others,[106] have previously tested compound **2** (Table 3) and its corresponding Zn(II) and Ni(II) complexes (Figure 11) for their ability to inhibit HIV-1(III$_B$) replication in MT-4 cells and, indeed, all three analogs were effective. The transition metal complexes were also poor inhibitors of 12G5 binding to CXCR4 in a similar manner to compound **2**, suggesting a common mode of binding to CXCR4 that is somehow different to the way in which AMD3100 and its corresponding metal complexes bind.

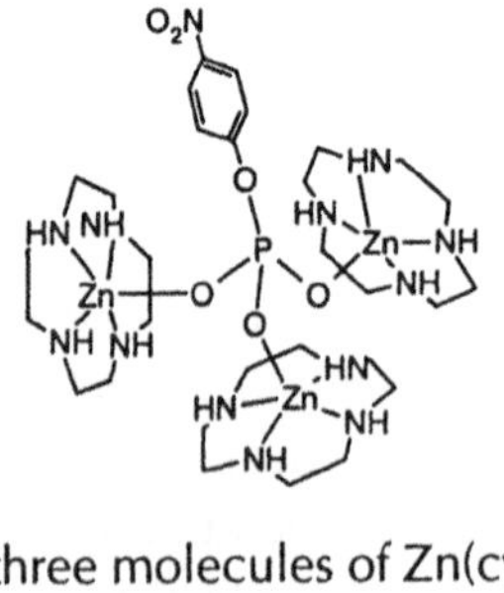

Figure 9. Complex of three molecules of Zn(cyclen) with a phosphate derivative. The dianionic charge of the complex was found to be equally distributed over the three zinc-bound oxygen atoms of the phosphate. Reprinted with permission from Kimura et al. *J. Am. Chem. Soc.* **1997**, *119*, 3068. Copyright 1997 by the American Chemical Society.

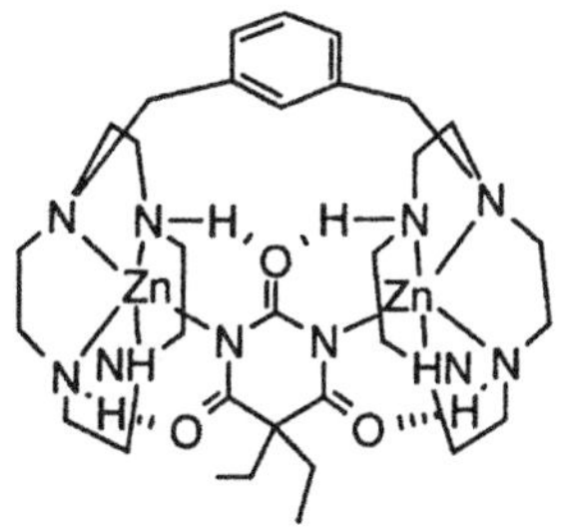

Figure 10. Proposed structure of a bis-zinc complex of compound **2** with barbital.

To complete the model, it remains necessary to explain the reduced antiviral activity of the Co and Pd complexes compared to the kineti-cally labile Zn, Ni, and (to a lesser extent) Cu complexes of AMD3100 (see Table 7). As the above examples show, the intermolecular inter-action of a carboxylate (or other ligating) moiety with a transition metal encapsulated within the cyclam rings of AMD3100 must be in an axial position of a metal complex. We therefore reviewed the propensity of Pd and Co complexes to form similar octahedral com-plexes. While the Pd(II) complexes with amine donors are rigidly square planar and do not form interactions with axial donors, the Co(III) complexes are octahedral with a predominantly *trans*-III con-figuration of the macrocyclic ring.[107] In this case, the corresponding *cis*-isomer can be isolated because of the high kinetic stability of cobalt(III)(cyclam) complexes.[108-112] However, the *cis* configuration

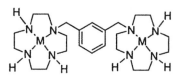

CPD#	EC$_{50}$ (μM) HIV-1(III$_B$)
2	0.07
M=Zn	0.19
M=Ni	0.03

Figure 11.

is unlikely to be the principal isomeric form of the bis-Co(III) complex of AMD3100 during antiviral testing due to the propensity of *cis*-cobalt(III)(cyclam) complexes to undergo isomerization in aqueous solution, affording the more thermodynamically stable *trans*-isomer.[111–113] Consistent with our previous observations, therefore, we suggest the reduced antiviral activity exhibited by the kinetically stable bis-Co(III) complex of AMD3100 (compound **23**, Table 7) is due to slower exchange of ligands at the axial positions. Consequently, a higher effective concentration of this compound is required to bind and inhibit HIV entry via CXCR4. Although we cannot state with any degree of certainty that the aforementioned examples prove that both AMD3100 and transition metal complexes of AMD3100 interact with amino acid residues containing, for example, carboxylate groups (Asp, Glu) within the second extracellular loop of CXCR4,[7] they do nevertheless demonstrate that these seemingly disparate structural types are able to participate in binding interactions with a common partner.

The role of transition metal complexation during binding to CXCR4 at the HIV inhibitory step was also further complicated by a recent study from our laboratories on bicyclam analogs containing a heteroaromatic linker. The HIV-1 and HIV-2 activity of pyridine and pyrazine linked bicyclams was found to be highly dependent upon the substitution of the heteroaromatic group connecting the cyclam rings[114] (Table 8). Whereas compounds containing a 2,6- or 3,5-substituted pyridine linker (compounds **25** and **27**) exhibited anti-HIV activities which were comparable to their phenyl counterpart **8**, compounds containing a 2,5- or 2,4-substituted pyridine group (**26**, **28**) exhibited substantially reduced anti-HIV activity (Table 8).

We have previously proposed a model for the deleterious effects of pyridine linkers in certain substitution patterns based on the ability of the pyridine-N group to participate in pendant complexation (the pseudo-axial positions of an octahedral complex) with an adjacent macrocyclic ring. This interaction may involve hydrogen bonding to a proton already complexed within the cyclam ring or coordination to a transition metal complex formed *in situ*. Presumably, when pendant conformations can occur, the resulting complex presents the wrong molecular shape for binding to CXCR4, irrespective of the presence or absence of transition metal ions.

Table 8.

CPD#	Linker	EC$_{50}$ (μM)		CC$_{50}$ (μM)
		HIV-1	HIV-2	
7		0.0042	0.0059	> 421
8		0.0337	0.0442	> 421
25		0.0245	0.0654	> 409
26		0.9081	2.1947	> 18
27		0.0316	0.0710	> 395
28		16.367	16.367	> 17

Our conformational hypothesis was deduced from a study of the coordination chemistry literature. Azamacrocycles such as cyclam display a rich coordination chemistry, forming complexes with a broad range of metal ions with high thermodynamic stability.[115] However, in order to fill the vacant coordination sites on the majority of transition metals complexed within the macrocyclic ring cavity, additional pendant coordinating groups have been attached to the periphery in order to fully encapsulate the metal ion. This has two major advantages: first, completing the coordination sphere of an octahedral complex by increasing the ligating ability of the macrocyclic ring can increase the thermodynamic stability of the resulting complex; and second, in some cases, the pendant coordinating arm can assist the rate of metal complexation by prechelation of the metal ion prior to the relatively slow step of transfer into the macrocyclic cavity. The chemistry and structure of pendant-substituted azamacrocycles and their transition metal complexes has been the subject of several reviews and will not be covered here in great detail.[114,116,117] However, if one reviews the anti-HIV activity of the 2,4- and 2,5-substituted pyridine-linked bicyclams analogs (28, 26) in context of their pendant complexation capabilities, it would appear that the pyridine nitrogen in these analogs is amenable to pendant interactions with the adjacent macrocyclic ring without interference from the second cyclam ring or the corresponding methylene group to which it is connected (Figure 12). In contrast, the 3,5-substituted pyridine group (compound 27) inherently precludes the formation of pendant conformers, and the pyridine-N of the 2,6-disubstituted pyridine analog 25 is inaccessible due to the steric requirements of the substituents (2,6-lutidine is generally used as a non-nucleophilic base). We subsequently found that the structure–activity relationships exhibited by pyridine-linked bicyclams are not unique: both pyrazine- and aniline-linked analogs displayed anti-HIV activities consistent with the pendant coordinating rationale.[114]

In order to test our hypothesis that potentially pendant-coordinating functional groups in the linker have a deleterious effect on the anti-HIV activity of bicyclams, we reasoned that incorporating a sterically

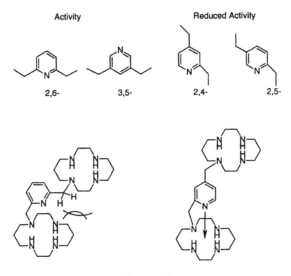

Figure 12.

demanding group at the position 6 of the 2,4-substituted pyridine linked bicyclam analog (**28**) should prevent pendant conformations from occurring to give an analog with comparable activity to the *meta*-phenylene-linked analog **8**. For this purpose, we chose to synthesize a 6-phenylpyridine-linked bicyclam (Figure 13). Examination of molecular models suggested that the pyridine-N in this compound was, to all intents and purposes, completely precluded from participating in pendant-coordinating interactions in a position perpendicular to the plane of the azamacrocyclic ring ring (the pseudo-axial positions of an octahedral complex), due to the steric bulk of the phenyl ring. The 6-phenyl pyridine bicyclam analog indeed inhibited HIV-1(III$_B$) and HIV-2 (ROD) replication in MT-4 cells with EC$_{50}$'s of 0.04 and 0.02 μM, respectively (compare to **8**), thereby confirming our supposition. As we will demonstrate however, the nature of the pendant interaction can be reasonably explained by intramolecular hydrogen bonding to the protonated cyclam ring or coordination to a transition metal complex.

Kimura et al. have shown that when a pyridine group is appended to the periphery of a cyclam ring in a manner suitable for pendant complexation, the pK_a of the pyridine group is substantially reduced.[118-120] This property has been suggested to indicate the presence of an intramolecular hydrogen bond between the pyridine-N and a shared proton already encapsulated within the macrocyclic ring.[120] Although the mutually repulsive effects of adding a third proton in close proximity to the protonated cyclam ring may also explain the reduced basicity of the pendant pyridine (an observation which also

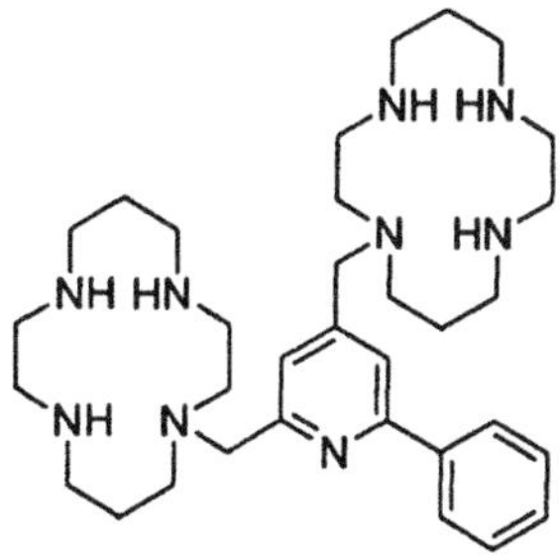

Figure 13. EC$_{50}$ against HIV-1 (III$_B$) = 0.04 μM.

explains the reduced pK_a's of the third and fourth secondary amines of cyclam), we nevertheless explored this suggestion in more detail with our system. To this end, we prepared the pendant-coordinating portion of the appropriate bicyclam analog (compound 28, Table 8) and attempted to measure the pK_a of the pyridine group (Figure 14). In titration and UV absorption experiments, the pK_a of the pyridine group could only be estimated to be less than 3, a value that is at least two log units lower than pyridine ($pK_a = 5.25$) and indistinguishable from pK_{a3} of cyclam.[114] Thus, the protonation constant of the pyridine moiety is entirely consistent with the hydrogen-bonding proposal of Kimura.

The transition metal complexation properties were also investigated (Figure 14). Reaction with one equivalent of nickel(II) diperchlorate in n-butanol gave a bright purple precipitate which exhibited a chemical composition corresponding to the desired nickel complex (Figure 14). The UV spectrum of this compound in aqueous 1M $NaClO_4$ (pH 6.9) gave three bands [327 (sh), 531 ($\varepsilon = 10$), and 934 ($\varepsilon = 9.1$) nm] characteristic of five-coordinate, high-spin Ni(II) complexes in which the pyridine group is pendant-coordinated.[121–128] We also found that coordination of the pyridine group was reversible due to pH-dependent competition of the nickel cation and a proton for the pyridine nitrogen: dissolving the purple solid in 5 M $HClO_4$ gave an orange solution with a single band at 462 ($\varepsilon = 40$) nm in the UV spectrum, a characteristic chromophore of low-spin, square-planar Ni(II) tetraamine complexes. Amusingly, alternate additions of sodium hydroxide and perchloric acid to this solution shuttle the equi-

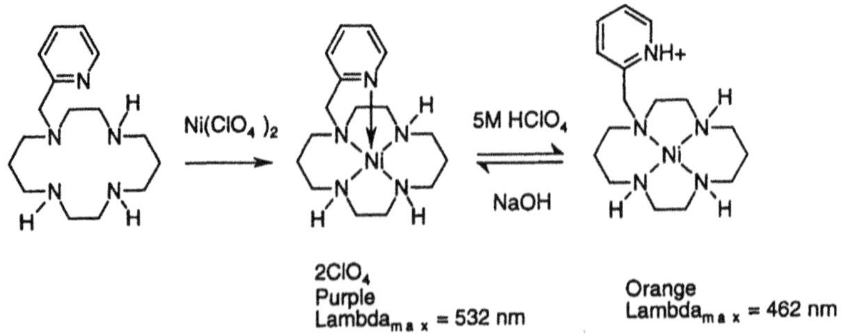

Figure 14. Reaction of the pendant coordinating portion of the pyridine linked bicyclam (28) with nickel perchlorate.

librium from the nonpendant (orange) to the pendant (purple) coordinated complex and *vice versa* (Figure 14).[121,122,124] These combined results provide unequivocal evidence that the pyridine group in Figure 14, and by analogy, the pyridine-linked bicyclam analogs **26** and **28** which exhibit reduced anti-HIV activity, are able to adopt pendant-coordinating conformations.

To complete our studies, we were faced with the inevitable question of whether the introduction of the phenyl substituent at position 6 of the pyridine ring actually prevents the formation of a pendant-coordinated complex upon chemical reaction with a transition metal salt. For this purpose, the analogous pendant-coordinating portion of the compound in Figure 13 was prepared and reacted with one equivalent of $Ni(ClO_4)_2$. Under these conditions we were gratified to observe the formation of an orange (rather than purple) colored complex (Figure 15) whose UV absorption spectrum (in acetonitrile) exhibited a single band at 468 ($\varepsilon = 23$) nm, again consistent with the formation of a low-spin, square-planar Ni(II) complex, but, in this case, the high-spin chromophore was absent.

Clearly, and in direct contrast to the compound in Figure 14, the phenyl group of the *N*-pyridyl cyclam in Figure 15 prevents approach of the pyridine-N in the pseudo-axial position of the adjacent nickel cation and therefore precludes formation of a pendant, five-coordinate complex. Though we were unable to definitively prove the involvement (or lack of involvement) of metal ions in the anti-HIV activity of bicyclams, the reduced activity of pyridine-linked bicyclams could be directly assigned to the proximal pendant pyridine moiety.

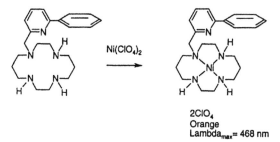

Figure 15. Reaction of the pendant coordinating portion of the compound in Figure 13 with nickel perchlorate.

Taking into account the lessons we had learned from substitutions in the linker, we refocused on our primary goal—identification of a potent antiviral agent which is orally absorbed. In this regard, the eight amine groups of AMD3100 were considered to be the dominant structural impediment for efficient transport across the gastrointestinal tract for two reasons: first and foremost, the cyclam rings are protonated twice at physiological pH and, second, the protonated rings possess multiple hydrogen-bond donors and acceptors. One approach we have pursued to overcome this difficulty is to define the minimum number of nitrogen atoms per ring required for potent antiviral activity. We reasoned that sequential replacement of the secondary amines with neutral heteroatoms or heteroaromatic groups (of lower pK_a than the secondary amines of cyclam) may retain the required macrocyclic ring conformation required for potent antiviral activity and simultaneously reduce the overall charge and hydrogen-bonding character.[129] For synthetic expediency, we initially prepared *para*-phenylene-bis(methylene) linked analogs of the [*iso*-14]aneN$_4$ macrocycle (*iso*-cyclam) in which a single secondary amine was replaced with an oxygen atom or a methylene group, and analogs of the [14]aneN$_4$ (cyclam) ring containing two oxygen or sulfur atom replacements. In antiviral testing (Table 9) the bis-[14]aneN$_3$O (**29**) and [14]aneN$_3$ (**30**) analogs exhibited substantially reduced anti-HIV-1 and anti-HIV-2 potency compared to the bis-*iso*-cyclam derivative **5**, consistent with

Table 9.

CPD#	Structure	EC_{50} (μM) HIV-1 (III$_B$)	IC_{50} (μM) 12G5 mAb Inhibition
29 (X = O)		4.49	NT
30 (X = CH$_2$)		9.61	>35
5 (X = NH)		0.03	0.067
31 (X = O)		>360	>35
32 (X = S)		6.8	>35
7 (X = NH)		0.004	0.007

their lack of ability to inhibit binding of the CXCR4-specific mono-clonal antibody 12G5 to CXCR4. The corresponding bis-[14]aneN$_2$S$_2$ (**32**) and [14]aneN$_2$O$_2$ (**31**) analogs were ineffective antiviral agents, displaying EC$_{50}$'s for inhibition of HIV-1 replication that were equal to their respective 50% cytotoxic concentrations in MT-4 cells. Clearly, the azamacrocyclic ring requires four nitrogen atom do-nors/acceptors for high binding to the molecular target on the CXCR4 receptor. We therefore shifted our attention to a series of compounds which contain heteroaromatic groups such as pyridine incorporated into the azamacrocyclic ring.

Recently, Costa and Delgado[130] have shown that 12- to 14-mem-bered tetraazamacrocycles containing pyridine have a lower pK_{a1} and pK_{a2} than the secondary amine groups of the cyclam ring and, further-more, the lowest pK_{a2} is exhibited by the macrocycle with the smallest number of ring members (py[12]aneN$_4$) as shown in Figure 16. These observations may not be entirely unexpected since the pyridine group should impart lower basicity to the macrocyclic ring and, as we have previously mentioned, the degree of ring protonation is controlled by the coulombic repulsion of confining multiple positive charges in a limited three-dimensional space. However, the results strongly sug-gest that if the pK_{a2} of the macrocyclic ring could be lowered to 7.4 or less without compromising antiviral activity of the subsequent dimer, a significant fraction of the individual macrocyclic rings would exhibit an overall charge of +1 at physiological pH, rather than the charge of +2 displayed by the cyclam ring.

Although the pK_{a2} of the py[14]aneN$_4$ macrocycle (pK_{a2} = 8.56) is still about 1.2 log units higher than physiological pH (Figure 16), the

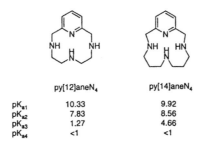

	py[12]aneN$_4$	py[14]aneN$_4$
pK_{a1}	10.33	9.92
pK_{a2}	7.83	8.56
pK_{a3}	1.27	4.66
pK_{a4}	<1	<1

Figure 16.

inherent pK_a of the pyridine group can be easily adjusted by incorporating electron-withdrawing (lower basicity) or electron-donating
(higher basicity) groups in position 4. In principle, these structural
modifications alone could significantly improve gastrointestinal absorption.

To test our hypothesis, we initially prepared, and evaluated the
antiviral efficacy of an analog containing py[14]aneN$_4$ rings connected via a *para*-phenylenebis(methylene) linker (Table 10). Compound **33** inhibited HIV-1 replication in MT-4 cells with an EC$_{50}$ of
0.5 µM while remaining nontoxic at concentrations exceeding 200
µM. Although the antiviral potency of **33** was lower than that of
AMD3100, we subsequently synthesized a series of compounds in
which the pK_a of the heteroaromatic group was systematically
varied. The 4-chloropyridine-substituted bis-macrocycle **34** exhibited comparable activity to **33** but was significantly more cytotoxic
(the CC$_{50}$ of **34** was 18 µM). The introduction of electron-donating
groups such as methoxy (**36**) or phenyl (**37**), however, reduced antiviral activity: the EC$_{50}$'s for analogs **36** and **37** were approximately
threefold and fivefold lower than the concentration of **33** required to
inhibit HIV-1 replication by 50%. One could conclude, therefore, that
in this series of compounds, the pK_a of the pyridine group can be
lowered without affecting the antiviral activity. To test this deduction
further, we reasoned that incorporation of a pyrazine (pK_a = 0.65)
rather than pyridine (pK_a = 5.2) group into the macrocyclic ring should

Table 10.

CPD#	X	EC$_{50}$ (µM) HIV-1 (III$_B$)	CC$_{50}$ (µM)
33	H	0.53	> 199
34	Cl	0.75	18
35	OH	> 174	> 200
36	OMe	3.10	> 216
37	Ph	1.75	19

render the heteroaromatic group resistant to protonation but retain anti-HIV activity comparable to **33** and **34**. Inexplicably, the pyrazine analog was significantly less potent at inhibition of HIV-1 and HIV-2 replication, displaying EC_{50}'s of 30 μM and 20 μM, respectively.[129]

Having established that an unsubstituted pyridine group incorporated into the azamacrocyclic ring has the most favorable basicity profile for potent antiviral activity, a series of compounds were prepared in which the macrocyclic ring contained 12 and 16 ring members[129] (Table 11). In contrast to the structure–activity relationship observed in the series of aliphatic tetraazamacrocyclic compounds (Table 3), the *para*-phenylenebis(methylene)-linked dimers of the py[12]aneN$_4$ (**38**), py[14]aneN$_4$ (**33**, Table 10), and py[16]aneN$_4$ (**41**) ring systems exhibited comparable 50% effective concentrations for inhibition of HIV-1 (and HIV-2) replication (Table 11). The corresponding *meta*-phenylenebis(methylene)-linked dimer of the py[12]aneN$_4$ macrocycle (**39**) displayed higher antiviral potency for inhibition of HIV-1 and HIV-2 replication than the corresponding *para*-analog **38** (and the activity against HIV-2 was greater than against HIV-1), consistent with the structure–activity relationship of the aliphatic tetraazamacrocycles (compounds **1** and **2**, Table 3) previously reported. To complete the study, a *para*-phenylenebis(methylene)-linked dimer of the py[*iso*-14]aneN$_4$ macrocycle was prepared. An isomer of the py[14]aneN$_4$ macrocyclic ring of analog

Table 11.

CPD#		Structure	EC$_{50}$ (μM)		CC$_{50}$ (μM)
			HIV-1 (III$_B$)	HIV-2 (ROD)	
38	*para*		0.5238	0.1273	8.5
39	*meta*		0.0971	0.0025	30
40	$n = 1$		0.0008	0.0016	194
41	$n = 2$		0.4213	1.2068	142

33, AMD3329 (**40**), exhibited an EC_{50} against HIV-1 replication that was 3 orders of magnitude lower than the EC_{50} of **33**: the 50% effective concentrations of **40** against HIV-1 and HIV-2 replication were 0.8 nM and 1.6 nM; the CC_{50} in MT-4 cells was 199 μM, which gives a selectivity index for AMD3329 (**40**) against HIV-1 of greater than 2.4 × 10^5. In addition, AMD3329 (**40**) inhibited HIV-1 and HIV-2 replication at a 3–5-fold lower concentration than the concentration of AMD3100[7] required to inhibit viral replication by 50%.

AMD3329 (**40**) and AMD3100 (**7**) were also compared for their ability to interfere with virus-induced syncytium formation between persistently HIV-1(III_B)-infected HUT-78 cells and uninfected MOLT-4 cells. In these assays, the concentration of AMD3329 re-

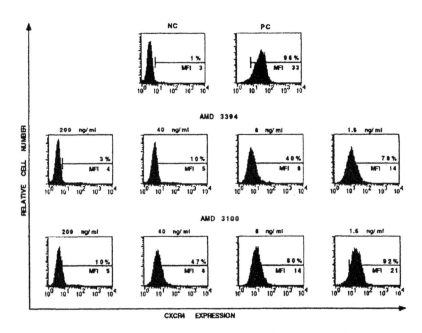

Figure 17. Inhibition of the binding of an anti-CXCR4 mAb (12G5) to SUP-T1 cells in the presence of AMD3394 (AMD3329 isolated as the octahydrochloride salt) and AMD3100 at different concentrations.[129] NC is the isotype control mAb and PC is the specific staining obtained with 12G5 in the absence of test compound. The percentage of positive cells and the MFI values are indicated in each histogram.

quired to inhibit syncytium formation by 50% was 90-fold lower than that of AMD3100 (EC_{50}'s for AMD3329 and AMD3100 were 0.012 and 1.130 μM, respectively).

In order to confirm that AMD3329 was mechanistically equivalent to AMD3100, their interactions with CXCR4, the main coreceptor for entry of T-tropic HIV strains, were directly compared as shown in Figures 17 and 18. Both AMD3100 and AMD3394 (the octahydro-chloride salt of AMD3329, see Figure 19) dose-dependently inhibited binding of the CXCR4-specific monoclonal antibody 12G5 to CXCR4 with IC_{50} values of 0.007 and 0.001 μM, respectively (Figure 17), and blocked SDF-1α-induced Ca^{2+} flux ($[Ca^{2+}]_i$) in SUP-T1 cells: AMD3394 completely blocked the $[Ca^{2+}]_i$ increases induced by SDF-1α at 7 nM, whereas at this concentration AMD3100 partially inhibited (42%) the $[Ca^{2+}]_i$ increase and completely blocked the response at 120 nM[5] (Figure 18). These combined assays unequivocally dem-

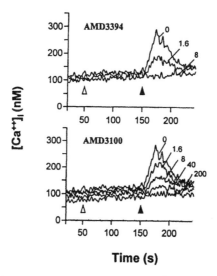

Figure 18. Inhibition of SDF-1α induced Ca^{2+} flux in SUP-T1 cells by AMD3394 (the octahydrochloride salt of AMD3329) and AMD3100 at various concentrations (shown in ng/mL).[129] Test compounds or buffer were added at 50 s (Δ) and SDF-1α was given as a second stimulus at a concentration of 30 ng/mL (▲), 100 s after addition of the test compound.

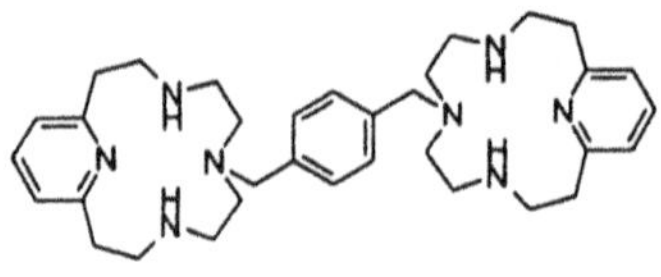

Figure 19.   Structure of AMD3329/AMD3394.

onstrate the more potent interaction of AMD3329 (AMD3394) with CXCR4, consistent with its higher antiviral activity.[129]

V. SYNTHESIS OF BIS-AZAMACROCYCLIC ANTIVIRAL AGENTS: NITROGEN-LINKED BIS-AZAMACROCYCLES

The chemistry of azamacrocyclic systems has been the subject of several substantial reviews and will not be covered here in any detail.[131,132] However, this review will profile the approaches we have taken to prepare macrocycles of varying ring size and functionality and briefly review the latest developments in the preparation of azamacrocycles pertinent to the synthesis of bis-macrocyclic antiviral agents.

With few exceptions described herein, the vast majority of antiviral bis-azamacrocycles we have reported to date are connected to the linker via a nitrogen group on the azamacrocyclic ring. At first glance this would appear to be a relatively trivial pursuit. Several groups have prepared bis-azamacrocyclic compounds by the general route shown in Scheme 1 in which a tris-N-protected tetra-azamacrocycle, containing a single free secondary amine group, is reacted with one-half an

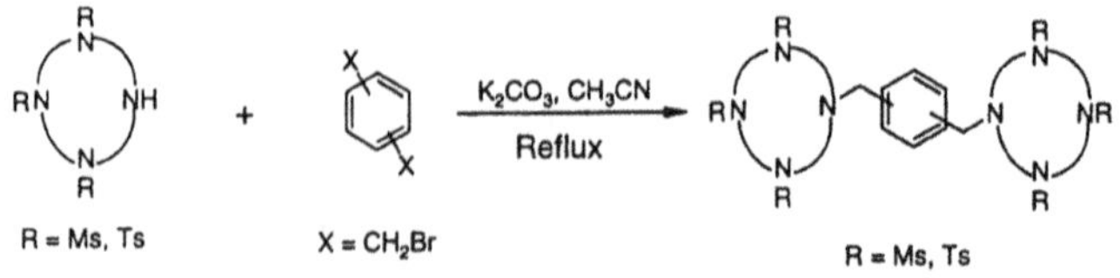

Scheme 1.

equivalent of an appropriate bis-electrophile, and the subsequent intermediate is deprotected to give the desired bis-macrocycle.

In effect, the two steps described above are common to virtually every N-linked bis-azamacrocycle described in this chapter. However, depending upon the ring structure, it is the construction of a suitably (protected) azamacrocycle precursor which poses the greatest synthetic challenge and these can be divided into two basic approaches: synthetic manipulation of commercially available azamacrocycles and regioselective construction of azamacrocycles. Examples of syntheses from both classes will be described in more detail.

A. Synthesis of Bis-azamacrocycles from Azamacrocyclic Rings

The preparation of tris-N-protected derivatives of cyclam ([14]aneN$_4$) and cyclen ([12]aneN$_4$) are relatively straightforward due to the predictable regiochemical outcome of the protection reaction. For example, reaction of cyclam with 1.8–2.1 equiv of toluenesulfonyl chloride (Ts or tosyl) according to the procedure of Fabrizzi gives tris-N-Ts-cyclam in 30–40% yield after recrystallization[133] (Scheme 2). If the amount of toluenesulfonyl chloride is increased in this reaction (closer to the theoretical stoichiometry of 3 equiv) the major product isolated is tetra-Ts cyclam. Presumably, this is due to intramolecular activation of the unreacted amine groups via hydrogen-bonding interactions. In contrast, reaction of cyclen with 3 equiv of Ts–Cl gives tris-Ts cyclen in a 70–80% isolated (Scheme 2).

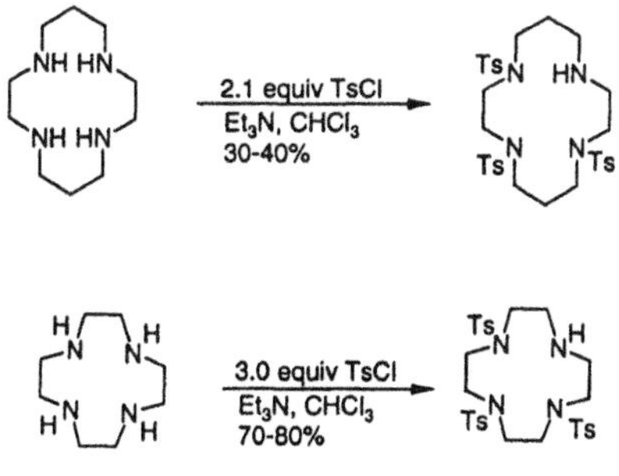

Scheme 2.

These precursors were used to prepare a variety of alkyl-linked and phenylenebis(methylene)-linked tetraazamacrocyclic dimers[89,114] using modified literature procedures[133] as shown in Scheme 3. In general, our deprotection method of choice for removal of toluenesulfonamido protecting groups is hydrolysis with 48% aqueous hydrobromic acid and acetic acid at reflux, due to the occasional fortuitous precipitation of the desired bis-tetraazamacrocycle as the octahydrobromide salt. Nevertheless, these harsh deprotection conditions can be problematic, depending upon the functional groups within the linker and the number of ring members. For example, whereas deprotection of tosyl groups on alkyl- or phenylenebis(methylene)-linked dimers of the cyclam ring (14-ring members) provide high yields of bicyclams, the tosyl groups of aromatic-linked bis-cyclens (12-membered) could not be deprotected without significant cleavage of the macrocyclic ring(s) from the dimer at the benzylic position. Alternative attempts to deprotect the tosyl groups under acid conditions (concentrated H_2SO_4, 100 °C, 2–3 hours) gave similar results. In this case, reductive cleavage of the tosyl groups with 2% sodium amalgam [Na(Hg)/THF/MeOH/Na_2HPO_4/reflux] provided the free base of the desired phenylenebis(methylene)-linked bis-cyclens which were converted to their corresponding hydrochloride or hydrobromide salts.

On several occasions, however, even aromatic-linked bicyclams proved too sensitive to the tosyl deprotection conditions for two reasons: (a) the substituents on the aromatic linker enhanced cleavage

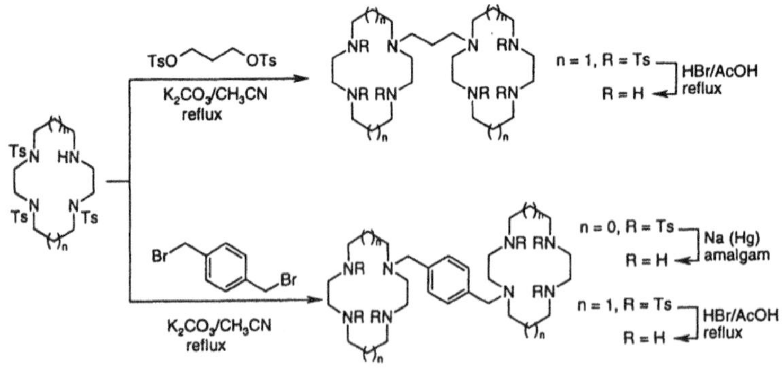

Scheme 3.

of the cyclam ring from the dimer at the benzylic position in a similar manner to cyclen, or (b) the aromatic ring substituent was unable to withstand acidic deprotection conditions. Therefore, an alternative protecting group was sought. For several reasons (which will become more apparent later in this chapter) we chose diethylphosphoramidate (Dep) protection, originally reported by Mertes et al.[134] The analogous reaction of cyclam with diethylchlorophosphate gave tris-Dep cyclam in a comparable yield to formation of tris-Ts cyclam and reaction of this intermediate with, for example, 2-phenyl-α,α'-dibromo-*p*-xylene or the analogous 2-carbomethoxy derivative, gave the fully protected bicyclams in which all secondary amines are protected with Dep[89,114] (Scheme 4). Unlike deprotection of the tosyl groups, which requires hydrobromic acid at reflux, the Dep groups were smoothly deprotected with a freshly prepared solution of HBr(g) in acetic acid at room temperature to give the desired bis-macrocycles as their octahydrobromide salts.

Although the routes described above allowed the synthesis of a wide variety of bicyclams and bicyclens with varying functionality, they are generally unsuitable for larger scale synthesis due to the uneconomical consumption of cyclam [the cost of cyclam (U.S. $45/gram) and the 30–40% yield of the primary protection step (Tosyl or Dep) leave considerable room for improvement]. To this end, several notable improvements in methodology were reported during the course of our work and will be briefly summarized.

Handel et al. recently reported a novel synthesis of cyclam from the corresponding butanedione protected linear tetraamine[135] (Scheme 5). Reaction of commercially available *N,N'*-bis(2-aminoethyl)-1,3-

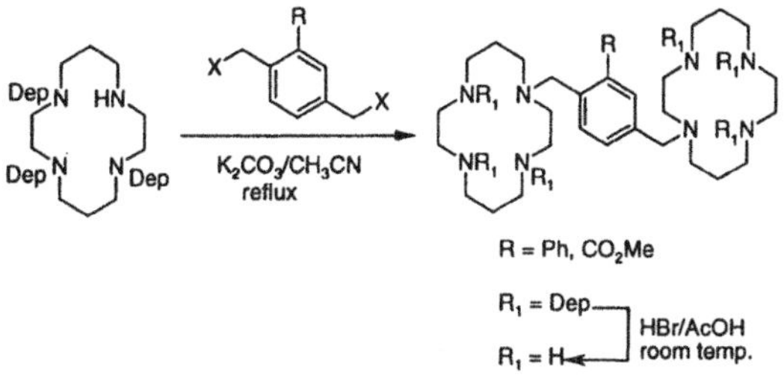

$$R = Ph, CO_2Me$$

$$R_1 = Dep$$
$$R_1 = H \qquad \text{HBr/AcOH room temp.}$$

Scheme 4.

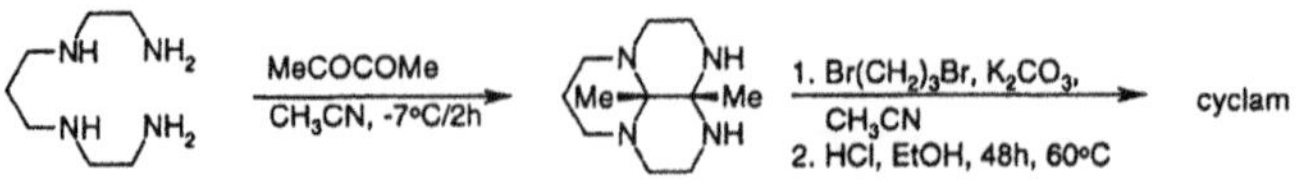

Scheme 5.

propanediamine with butanedione gave exclusively the tricyclic bis-aminal, in which the methyl groups are in a *cis* configuration and in 95% yield after recrystallization from hexane. The authors noted that in the *cis* configuration the two secondary nitrogen functions are correctly positioned for a (1+1) cyclo-condensation. Thus, treatment of the *cis*-dimethyl tricyclic bis-aminal with 1,3-dibromopropane gave the corresponding protected azamacrocycle in 90% yield, which was hydrolyzed under mild acidic conditions (dilute HCl solution) to quantitatively afford cyclam as the hydrochloride salt. This methodology has also been extended to the synthesis of cyclen and homocyclen.

A similar conceptual approach for the synthesis of cyclen was reported earlier by Weisman and Reed,[136] in which the key step of the synthesis is a double reductive ring expansion of the tricyclic bis-amidine with DIBALH (Scheme 6). S-alkylation of dithiooxamide with excess bromoethane followed by reaction of the resulting bis-thioimido ester salt with triethylenetetraamine afforded the tricyclic bis-amidine in 69% yield. Reduction with DIBALH in refluxing toluene followed by workup with NaF gave cyclen in 83% yield after sublimation. This methodology has also been extended to the synthesis of homocyclen[137] ([13]aneN$_4$).

The most efficient route to preparation of N-functionalized azamacrocycles would be to selectively react one of the nitrogen atoms without protecting the remaining amine groups of the ring. To this end,

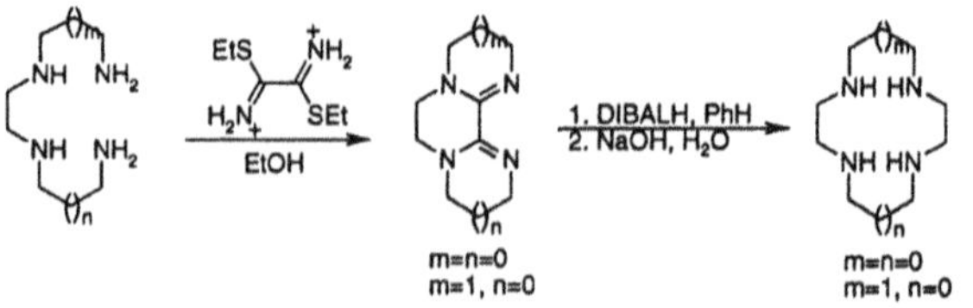

Scheme 6.

Kruper et al. reported the direct mono-N-alkylation of 12-, 14-, and 15-membered tetra-azamacrocycles using an electrophile/macrocycle ratio of unity.[138] Specifically, the reaction of cyclen with a variety of substituted α-halo esters in chloroform proceeded with excellent selectivity, affording the mono-alkylated products as the monohydrobromide salts in yields ranging from 70 to 80% after flash chromatography (Scheme 7). Small amounts of bis-alkylated macrocycle were also formed but the ratio of mono:bis was 40:1. The unexpected chemoselectivity was explained in terms of the strong affinity of the alkylated product for a single proton, resulting in diminished nucleophilicity of the remaining secondary nitrogen groups. Selectivity was found to be dependent upon a number of factors including macrocyclic ring size, solvent polarity, and the steric bulk of the electrophile. In view of these results, one can envisage a direct route to form bis-tetraazamacrocycles by reaction of a bis-electrophile with an unprotected azamacrocyclic ring. For example, Kimura et al. have prepared a bis-cyclen connected via a 1,3-phenylenebis(methylene) linker by direct reaction of α,α′-dibromo-*m*-xylene with cyclen. In this case, the selectivity for the mono-N-alkylated product was ensured by reaction of the bis-electrophile with a 6M excess of cyclen.[105] Optimization reactions to conserve the macrocycle by equalizing the stoichiometry were not performed.

The synthesis of bis-tetraazamacrocycles from unprotected cyclam in a one-step procedure was also found to proceed in good yield when the reaction was conducted with bis-acrylamides in chloroform in the presence of 1 equiv of *p*-toluenesulfonic acid[139] (Scheme 8). The authors accounted for the high selectivity by suggesting that in the

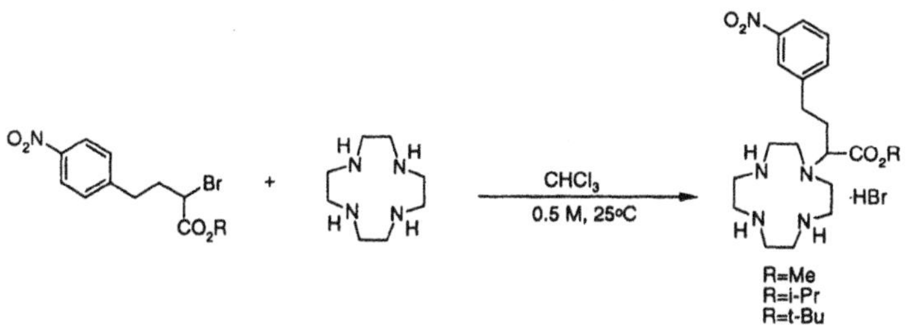

Scheme 7.

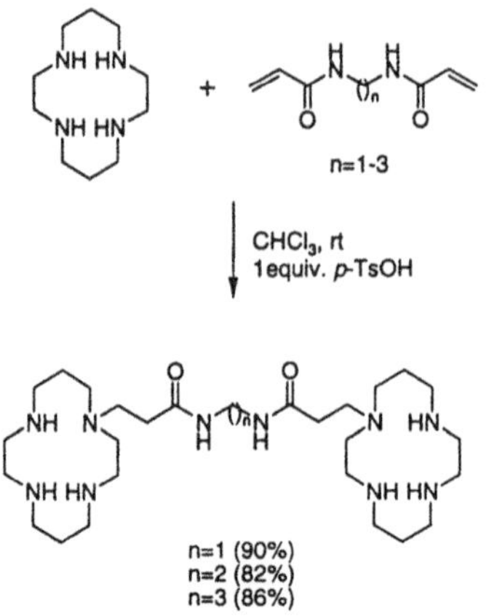

Scheme 8.

presence of 1 equiv of *p*-toluenesulfonic acid, cyclam is mono-pro-
tonated thus leaving only one nucleophilic nitrogen atom available for
Michael additions. It seems more likely, that the toluenesulfonic acid
is closely associated with the protonated cyclam ring in chloroform
solution, and the hydrogen-bonded complex is less susceptible to
further Michael additions.

B. Synthesis of Bis-azamacrocycles by Regioselective Construction of the Macrocyclic Ring

It does not take long to exhaust the library of commercially available azamacrocycles that are amenable to useful synthetic transformations. If one needs to prepare bis-azamacrocycles of a 15-membered
tetraazamacrocyclic ring for example, the individual ring is available,
but due to the reduction in symmetry compared to cyclam, there is
considerable ambiguity in the regioselectivity of reactions with the
free macrocycle. This is best illustrated by a structure/reactivity comparison of cyclam ([14]aneN$_4$) and isocyclam ([*iso*-14]aneN$_4$) in a
hypothetical reaction with benzyl bromide: whereas reaction of benzyl
bromide with cyclam can only lead to a single mono-N-alkylated
product, there are three possible regiochemical isomers which may
arise from a similar reaction with isocyclam (Scheme 9).

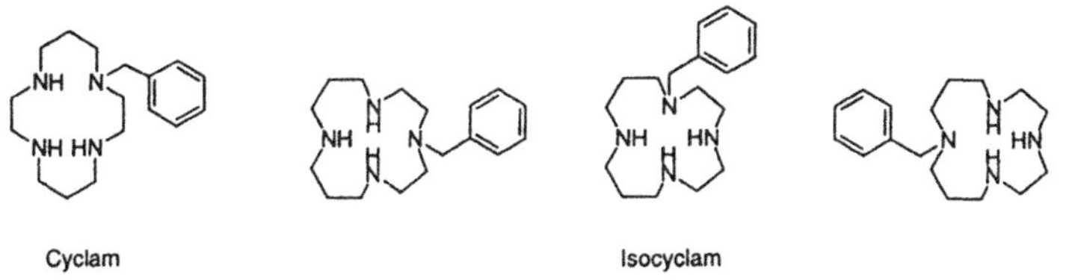

Scheme 9.

In order to prepare azamacrocycles of varying ring size, configuration, and functionality, therefore, it was necessary to develop a variety of routes to macrocyclic intermediates that ensure the regioselectivity of the reaction with the bis-electrophile during the dimerization step. To this end, we have (in many cases) used a combination of protecting groups for construction of the rings: (a) the toluenesulfonamido group which is required for the Richman–Atkins macrocyclization reaction,[140,141] and (b) a single, alternative protecting group targeted at the N-position where we ultimately intended to perform the dimerization reaction. A key requirement of the targeted group is the ability to be easily deprotected using conditions in which the tosyl groups are stable.

Historically, our first application of the strategy outlined above was the synthesis of the protected [*iso*-14]aneN$_4$ (isocyclam) macrocycle,[89] originally reported by Hediger and Kaden[142] (Scheme 10). The key intermediate for this synthetic route was the bis-toluenesulfonamide precursor which contained a benzyl group targeted for selective deprotection. The requisite tris-*p*-toluenesulfonate portion was obtained by tosylation of diethanolamine in CH$_2$Cl$_2$ in Et$_3$N. A Richman–Atkins cyclization of the two portions gave the benzyl-protected macrocycle, which was subjected to hydrogenolysis with Pd(OH)$_2$ in refluxing formic acid giving the desired tris-protected isocyclam intermediate.

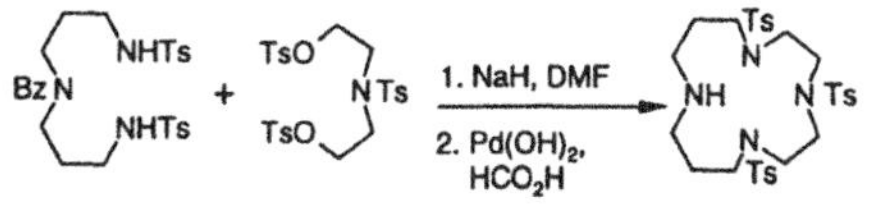

Scheme 10.

Alternatively, synthesis of the appropriately protected [15]aneN$_4$ macrocycle was accomplished by modification of the methodology developed by Mertes et al.[134] in which a diethoxyphosphoryl (Dep) group was targeted for selective deprotection (Scheme 11). A two-step derivatization of diethanolamine with 1.0 equiv of diethyl chlorophosphate (to give the phosphoramidate diol) followed by 2.0 equiv of methanesulfonyl chloride, under standard conditions afforded the Dep-protected portion in a straightforward manner. Macrocyclization with tris(p-tolylsulfonyl)-N-(2-aminoethyl)-1,3-propanediamine in the presence of excess Cs$_2$CO$_3$[143] (or K$_2$CO$_3$[144]) gave the required macrocycle in 55% yield after purification by column chromatography on silica gel. Finally, selective removal of the phosphoryl group with 30% HBr/acetic acid at room temperature gave the selectively tris-tosyl-protected [15]aneN$_4$. Various combinations of the routes described in Schemes 10 and 11 were used to prepare suitably tris-protected intermediates of the [13]aneN$_4$ and [16]aneN$_4$ macrocycles, respectively.[89]

Although cyclam is amenable to straightforward synthetic manipulation, the regioselectivity of reactions with cyclam derivatives can be problematic, particularly when substituents are introduced onto the periphery of the cyclam ring at a single carbon position. During the course of our work to determine the structural features of bicyclams required for potent anti-HIV activity, we wished to prepare cyclam analogs containing electron-donating and electron-withdrawing groups on the macrocyclic ring. The syntheses of the dimethyl- and fluoro-cyclam analogs shown in Schemes 12 and 13 provide complementary approaches to achieving this goal. The synthesis of the 6,6-dimethyl cyclam is shown in Scheme 12. Condensation of 2,2-dimethyl-1,3-propanediamine with chloroacetyl chloride afforded the corresponding bis-chloroamide. Macrocyclization of this intermediate with 1,3-diaminopropane gave the requisite dioxocyclam intermediate in which two of the amine groups are effectively "pro-

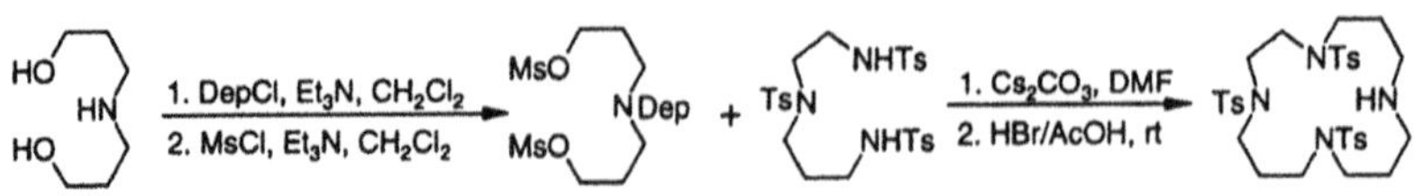

Scheme 11.

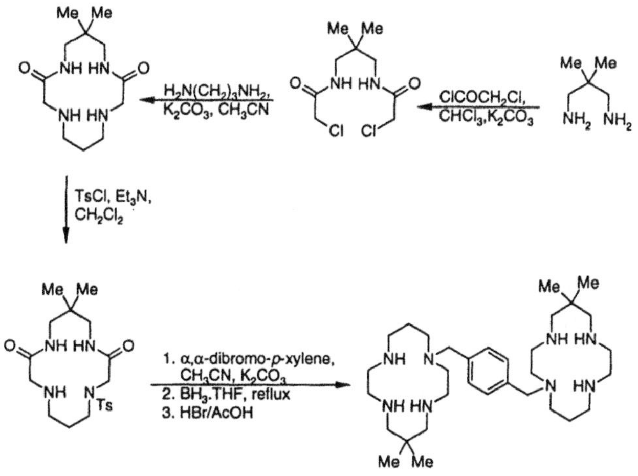

Scheme 12.

tected" as amides. In this case, one of the two remaining nitrogen groups was tosylated, leaving a secondary amine group for the dimerization reaction with α,α'-dibromo-*p*-xylene. To complete the synthesis, the amide groups were reduced with BH₃·THF and the tosyl-protecting groups removed by hydrolysis with HBr in acetic acid

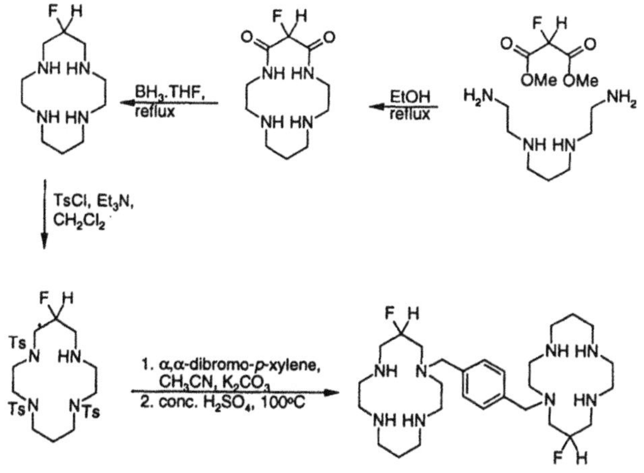

Scheme 13.

to give the desired bis-cyclam containing methyl groups at position 6 of both rings.

The 6-fluorocyclam ring system was prepared by the procedure of Kimura et al.[145] (Scheme 13). Condensation of N,N'-bis-(2-aminoethyl)-1,3-propanediamine with dimethyl fluoromalonate was carried out in refluxing ethanol to give the corresponding dioxocyclam. This intermediate can be used to prepare two possible regioisomeric bicyclam analogs, depending upon the required position of the fluoro group with respect to the amine group, to which the linker is connected. For example, the fluoro-substituted dioxocyclam could be viewed as an equivalent precursor to the dimethyldioxocyclam shown in Scheme 12, in which two amine groups are protected as amides. Similar functional group interconversions would lead to the corresponding p-phenylenebis(methylene)-linked bis-(6-fluoro)cyclam analog. In this case, the amide groups were reduced with $BH_3 \cdot THF$ to give the parent fluorocyclam (Scheme 13), then reacted with tosyl chloride and triethylamine to afford a single tris-protected fluorocyclam isomer. Presumably, the regioselectivity of this reaction is due to the electron-withdrawing effect of the fluoro group which reduces the nucleophilicity of the adjacent secondary amines. With the tris-tosyl derivative in hand, reaction with α,α'-dibromo-p-xylene, followed by deprotection with concentrated H_2SO_4 afforded the alternative fluoro-substituted bicyclam isomer (Scheme 13).

In some cases, the selective tosylation of a preconstructed azamacrocyclic ring is necessary but not always efficient. For example, in order to prepare a suitably protected macrocyclic precursor of the [14]aneN$_2$S$_2$ ring system, a cumbersome double-protection procedure was employed,[129] as shown in Scheme 14. The left-hand portion of the dithia ring system was prepared by reaction of ethanethiol and bromopropionitrile in CH_2Cl_2 in the presence of Et_3N to give the dinitrile in 88% yield. The dinitrile was then reduced to the diamine with $BH_3 \cdot THF$ and the intermediate amine was derivatized with p-toluenesulfonylchloride under standard conditions to give the ditosylamide. Macrocyclization was accomplished using standard conditions: dropwise addition of a DMF solution of ethyleneglycol ditosylate into a solution of the dithia-portion in DMF containing Cs_2CO_3 gave the fully protected [14]aneN$_2$S$_2$ macrocycle in a 42% isolated yield. To complete the synthesis of the symmetrical dimer, the

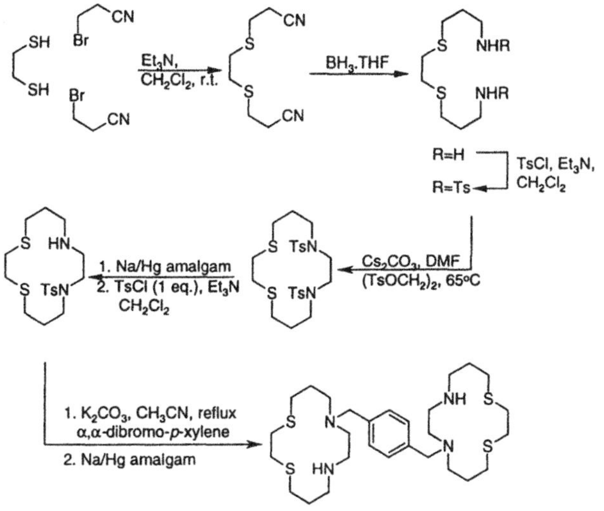

Scheme 14.

ditosyl-protected macrocycle was converted to the mono-tosylated intermediate by a two-step procedure involving deprotection of the tosyl groups with 3% sodium amalgam followed by selective protection of a single amine group. Dimerization of the available secondary amine with α,α'-dibromo-*p*-xylene in refluxing CH_3CN in the presence of K_2CO_3 gave the protected dimer, which was deprotected with 1% sodium amalgam and converted to the hydrochloride salt. In retrospect, while this approach accomplished our immediate goal, the efficiency of this particular route could be improved by macrocyclization of the unprotected diamine intermediate with chloroacetyl chloride that (theoretically) provides a suitably protected intermediate for the impending dimerization reaction in a single synthetic step.

Fortunately, azamacrocycles related to the [*iso*-14]aneN$_4$ ring system, featuring a single heteroatom substitution per macrocycle (N$_3$X macrocycles), are significantly more amenable to direct, regioselective synthesis using our preferred combination of diethoxyphosphoryl (Dep) and *p*-toluenesulfonyl (Ts) protecting groups.[129] The synthesis of the bis-[*iso*-14]aneN$_3$O macrocycle is shown in Scheme 15. The amino group of the dipropionitrile intermediate was first protected with diethylchlorophosphate in CH_2Cl_2 in the presence of Et_3N to give

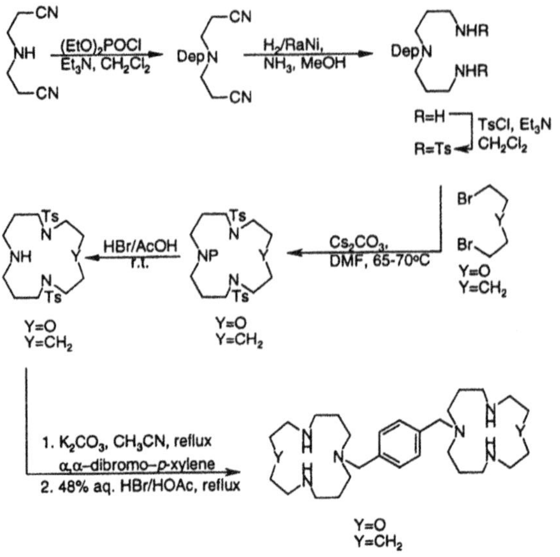

Scheme 15.

the Dep-protected dinitrile in 64% yield. Reduction of the nitrile groups by hydrogenation over Raney-Ni in a saturated solution of NH_3 in MeOH gave the diamine which was subsequently reacted with p-toluenesulfonylchloride under standard conditions to give the ditosylamide. Macrocyclization of the ditosylamide with 2-bromoethyl ether was highly efficient, affording the protected N_3O macrocycle in 79% yield after purification by silica gel chromatography. The secondary amine required for dimerization was readily liberated by deprotection of the Dep group with HBr/acetic acid at room temperature for 2–3 h to give the requisite macrocycle exhibiting a single secondary amine group. Conversion to the corresponding dimer was uneventful. By replacing 2-bromoethyl ether with 1,5-dibromopropane in the macrocyclization step we were also able to prepare a *para*-phenylenebis(methylene)-linked dimer of the [14]aneN$_3$ ring.

For the construction of azamacrocyclic rings containing an aromatic or heteroaromatic group (phenyl, pyridine, or pyrazine) within the macrocyclic framework, the Dep-protected ditosylamide from Scheme 15 was reacted with a series of bis-electrophiles whose syntheses are depicted in Scheme 16.[129] Compounds which contain a

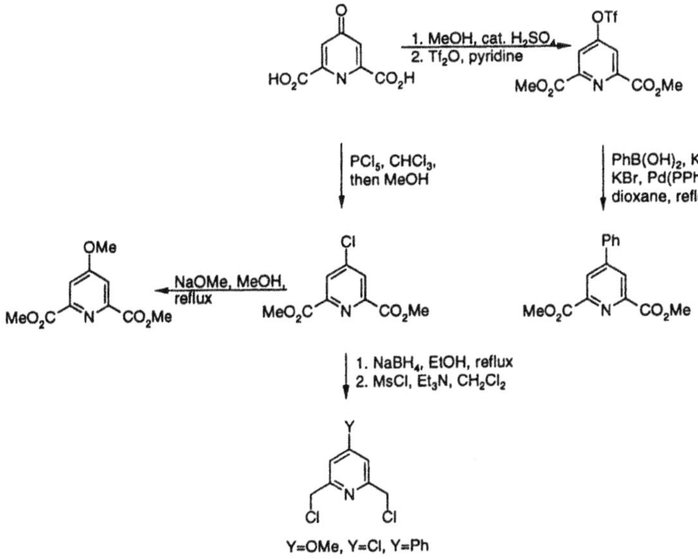

Scheme 16.

4-pyridine substituent, were synthesized from commercially available chelidamic acid (4-hydroxy-2,6-pyridinedicarboxylic acid). Thus, treatment of chelidamic acid with PCl_5 in refluxing $CHCl_3$, followed by quenching the intermediate acid chloride with methanol, afforded the corresponding 4-chloro-2,6-pyridinedicarboxylate dimethyl ester. Treatment of the diester with freshly prepared NaOMe in refluxing methanol gave the 4-methoxy derivative. The 4-phenyl derivative was also prepared from chelidamic acid. Esterification with MeOH and catalytic concentrated H_2SO_4 gave the corresponding dimethyl ester which was converted to the triflate by reaction with triflic anhydride in pyridine. Palladium-catalyzed cross-coupling of the triflate with phenylboronic acid in the presence of K_2CO_3 and KBr introduced the 4-phenyl substituent. The pyridine diesters were converted to bis-electrophiles in order to be utilized in the macrocyclization reaction: reduction with $NaBH_4$ in refluxing ethanol and subsequent reaction of the intermediate diols with methanesulfonyl chloride in CH_2Cl_2/Et_3N gave the equally suitable bis-chloromethyl derivatives rather than the expected bis-mesylate (presumably, by *in situ* nucleophilic substitution of the initially formed mesylate by chloride), even though the reaction conditions were mild (30 min at 0 °C).

Following macrocyclization and removal of the Dep group under standard conditions (HBr/acetic acid/room temperature/2–3 h) the series of macrocyclic precursors were used to prepare phenylene-bis(methylene)-linked dimers in a straightforward manner, exemplified by a dimer of the 4-methoxypyridine tetraazamacrocycle (Scheme 17). The synthesis of this analog also illustrates the versatility of the tosyl/Dep protecting group combination: the 4-methoxypyridine precursor can be used to prepare two analogs, depending on the choice of conditions used for the final deprotection step. Thus, treatment of the 4-methoxypyridine precursor with 48% aqueous HBr and acetic acid at reflux simultaneously cleaved the tosyl and methyl ether groups giving the 4-hydroxypyridine analog, whereas reaction with concentrated H_2SO_4 at 100–110 °C for 2 h selectively deprotected the tosyl groups in the presence of the methyl ether giving the 4-methoxypyridine analog.

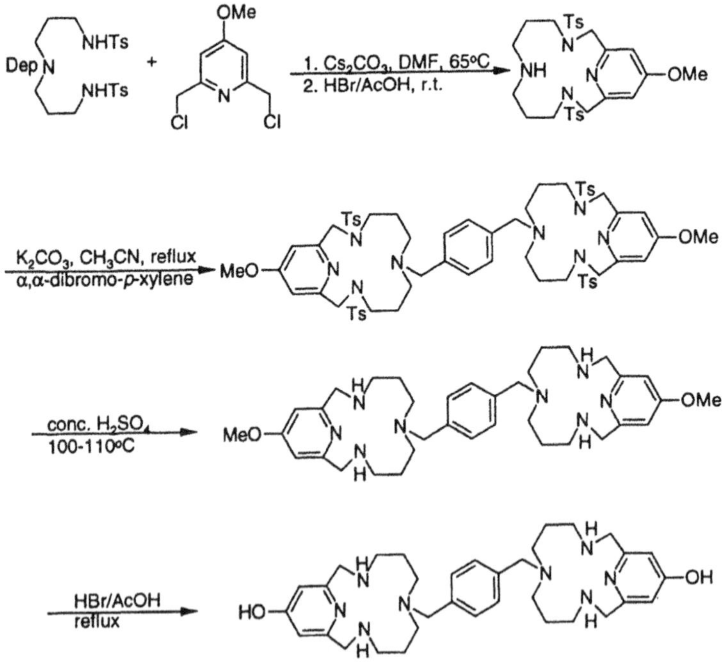

Scheme 17.

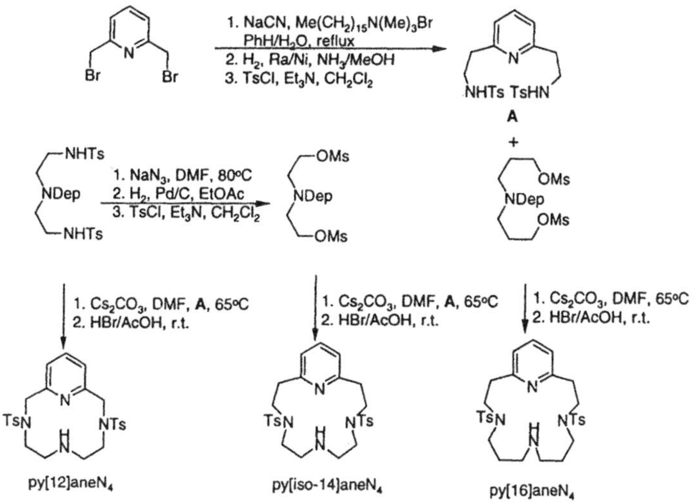

Scheme 18.

We have also extended the tosyl/Dep combination methodology to a second series of pyridine-containing tetraazamacrocyclic precursors which vary between 12 and 16 members per ring.[129] This was accomplished by manipulation of the macrocyclic precursors already in our possession. For example, suitably protected py[12]aneN$_4$, py[*iso*-14]aneN$_4$ and py[16]aneN$_4$ rings were obtained from a common intermediate, 2,6-bis(bromomethyl)pyridine as shown in Scheme 18. Macrocyclization with the appropriate bis-electrophiles containing a single N-Dep group afforded, post-deprotection, the requisite py[12]aneN$_4$, py[*iso*-14]aneN$_4$, and py[16]aneN$_4$ macrocycles, respectively. The corresponding dimers of the pyridyl macrocyclic rings were prepared as previously described and the intermediates were subsequently deprotected with HBr/acetic acid to give the octahydrobromide salts, respectively.

VI. SYNTHESIS OF BIS-AZAMACROCYCLIC ANTIVIRAL AGENTS: CARBON-LINKED BIS-AZAMACROCYCLES

Since the cyclam rings of our initial lead, AMD1657, were connected via carbon position 2, we decided to prepare analogs of AMD2763 and

AMD3100 in which the rings are connected at various carbon positions. There are several published strategies: the cyclam ring has been functionalized at carbon positions 5 and 6 by macrocyclization reactions of linear tetraamines with substituted α,β-unsaturated esters[145–147] and malonates.[147–151] The corresponding 2-carbon substituent can be introduced into the cyclam ring by macrocyclization of an ethylenediamine or amino acid synthon containing the desired C-substituent.[152–154] This approach has been predominantly used for the construction of C-functionalized [12]aneN$_4$ macrocycles[149,152] (see ref. 88 for related publications) in which the key macrocyclization step is the aminolysis of esters or reaction of deprotonated tosylamides with bis-electrophiles (the Richman–Atkins cyclization).

While we have successfully adapted the general strategies outlined above for the synthesis of C-linked bicyclams,[88,89] the synthetic efficiency of these approaches varies dramatically. For example, in order to prepare an analog of AMD3100 in which the cyclam rings are connected to the linker at the 6-carbon position, a strategy of malonate condensation with linear tetraamines was adopted, as outlined in Scheme 19. Reaction of α,α'-dibromo-p-xylene with 2.0 equiv of the sodium salt of diethyl malonate gave the requisite tetra-ester in a 40% isolated yield. A double macrocyclization reaction was then per-

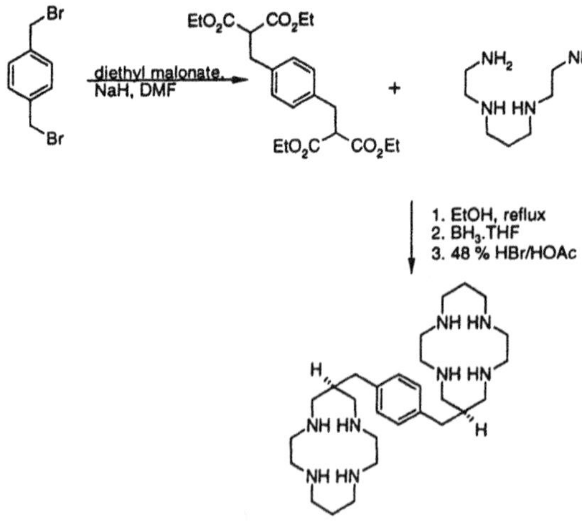

Scheme 19.

formed. Condensation of the tetra-ester with *N,N'*-bis(2-aminomethyl)-1,3-propanediamine in refluxing EtOH for 20 days affording the corresponding tetraamide in a highly unsatisfactory yield of 4%. The only redeeming feature of this reaction was the fortuitous precipitation of the pure product on day 19 of reaction. Reduction of the tetraamide with $BH_3 \cdot THF$ followed by aqueous HBr/acetic acid hydrolysis of the intermediate borane complex afforded the desired 6,6'-linked bicyclam analog in a 35% yield.

To prepare AMD2936, an analog of AMD2763 in which the cyclam rings are connected at carbon and nitrogen positions, we have developed new methodology based on the strategy of several groups.[155,156] From our perspective it was desirable to incorporate a versatile functional group directly upon the carbon backbone which was stable to the macrocyclization conditions, and was also capable of C–C bond-forming reactions on the fully constructed macrocyclic ring. Incorporation of a C-appended, *N*-methoxy-*N*-methylamide (Weinreb amide[157]) within a bis-toluenesulfonyl derivatized ethylenediamine synthon satisfied our criteria. The Weinreb's amide not only withstood the deprotonated tosylamide macrocyclization conditions (Richman–Atkins cyclization) but also served as a convenient carbonyl equivalent for further structural modification on the azamacrocyclic (cyclam) ring (for a review of the applications of *N*-methoxy-*N*-methylamides see ref. 158).

The key intermediate required for the synthesis of AMD2936 was the 2-(*N*-methoxy-*N*-methylcarboxamido)cyclam, which was convergently synthesized in six steps from commercially available amino acids as outlined in Scheme 20. The requisite Weinreb's amide, synthesized in three steps from 2,3-diaminopropionic acid, was macrocyclized with the appropriate dimesylate, synthesized from ethylenediaminedipropionic acid. Thus, dropwise addition of a DMF solution of the dimesylate into a solution of the Weinreb's amide in DMF (final concentration 0.022 M) containing excess Cs_2CO_3 maintained at a temperature of 65 °C gave the fully protected cyclam ring in a 70% isolated yield. Reduction of the amide with $LiAlH_4$ (1.5 equiv) in THF at −20 °C gave the aldehyde, which was subjected to a Wittig reaction with (carbethoxymethylene)triphenylphosphine (1.0 equiv) at room temperature in CH_2Cl_2 to give a 86:14 (*E:Z*) mixture of α,β-unsaturated esters in 64% overall yield. Stepwise reduction of

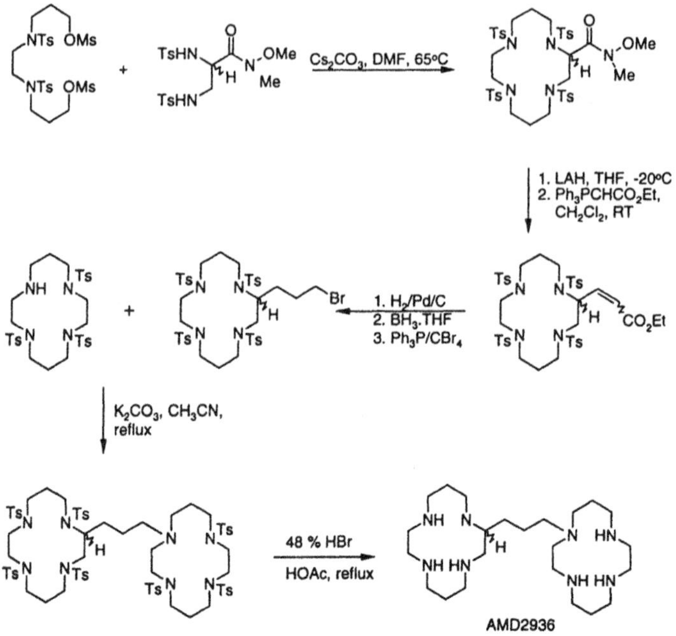

Scheme 20.

the esters with H_2/Pd/C (to give the saturated ester) followed by $BH_3 \cdot THF$ gave the corresponding alcohol which was subsequently converted to the aliphatic bromide in 85% yield by reaction with CBr_4/PPh$_3$ in refluxing benzene. To complete the synthesis of the desired bis-tetraazacyclotetradecane, the bromide was reacted with 4,8,11-tris(p- toluenesulfonyl)-1,4,8,11-tetraazacyclotetradecane in refluxing CH_3CN in the presence of excess K_2CO_3 to give the hepta-tosyl protected dimer in 62% yield after purification by column chromatography on silica gel. Finally, deprotection of the p-toluene-sulfonamido groups by hydrolysis with 48% aqueous HBr/acetic acid at reflux gave AMD2936, which precipitated from the reaction mix-ture as the octahydrobromide salt.

VII. SYNTHESIS OF AMD3100

In our laboratories, we have repeatedly prepared AMD2987 (the octahydrobromide salt of AMD3100) by our previously reported

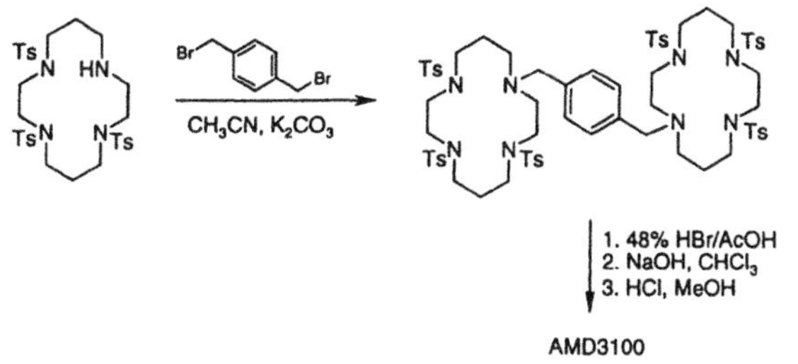

Scheme 21.

procedure[89] (Scheme 21). This material can be easily converted to the octahydrochloride salt (AMD3100) by conversion to the free base and treatment with anhydrous methanolic HCl. Several multikilogram syntheses of AMD3100 were completed via this route.

Recently, two alternative syntheses of AMD3100 were reported. The process development group at Sandoz reported a strategy based on selective N-protection of a tetraamine as the trifluoroacetamides[159] (Scheme 22). Thus, in a one pot sequence, the hexatosyl intermediate was prepared from N,N'-bis(3-aminopropyl)ethylenediamine by selective tris-trifluoroacetylation using 4 equiv of ethyl trifluoroacetate in THF followed by N-alkylation using 0.5 equiv of α α'-dibromo-*p*-xylene in the presence of diethylisopropylamine to afford the corresponding dimer. Hydrolysis of the trifluoroacetyl groups with NaOH

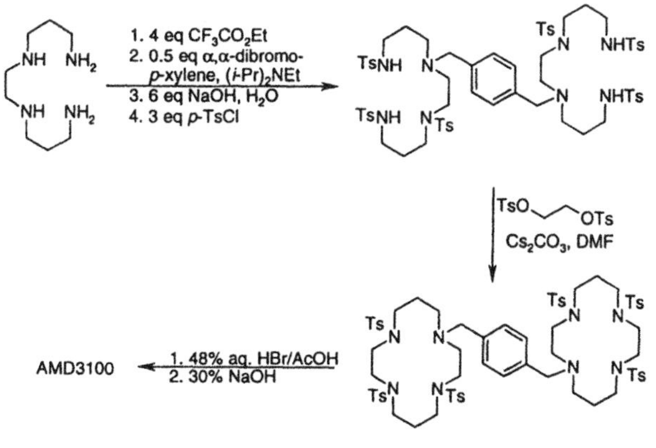

Scheme 22.

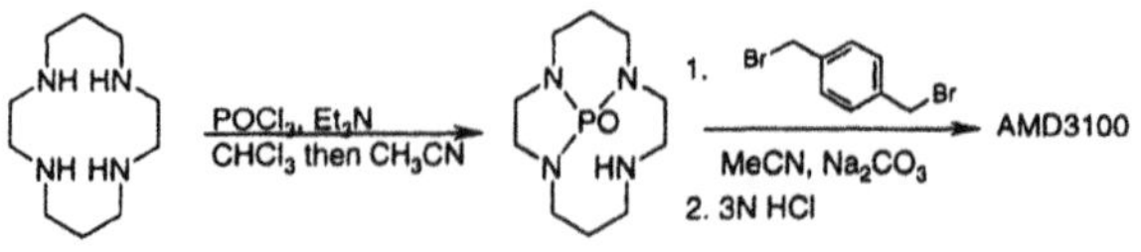

Scheme 23.

followed by reprotection with tosyl chloride afforded the hexatosyl intermediate in 80% yield after purification by column chromatography on silica gel. Treatment of the hexatosyl intermediate with ethylene glycol ditosylate in the presence of Cs_2CO_3 in DMF afforded the hexatosyl bismacrocycle in 35% yield after recrystallization from CH_3CN. Deprotection and conversion to the octahydrochloride salt was accomplished as previously described.

A second method for the synthesis of AMD3100 uses cyclam as a starting material and accomplishes selective N-protection via the phosphoric triamide[160] as outlined in Scheme 23. This approach was based on the method of intramolecular tris N-protection utilizing metal tricarbonyl complexes, trimethylsilyl, thiophosphoryl, boron, or phosphoryl moieties.[161] Again, in a one pot sequence, treatment of cyclam with 1 equiv of $POCl_3$ in the presence of a slight excess of triethylamine in chloroform (followed by acetonitrile) afforded the corresponding phosphoric triamide, which was N-alkylated with 0.5 equiv of α,α'-dibromo-*p*-xylene in the presence of Na_2CO_3 to afford the corresponding dimer. Removal of the protecting groups was accomplished by hydrolysis with aqueous HCl.

VIII. PHARMACOKINETICS OF AMD3100

In a Phase I clinical trial, AMD3100 was well-tolerated by healthy volunteers given a single parenteral dose as a 15 min intravenous infusion. Twelve subjects divided into three subjects per dose cohort received 10, 20, 40, 80 µg/kg of AMD3100. All subjects experienced a dose-related elevation in white blood cell counts (WBC's) of 1.5 to 3.5 times baseline values which returned to baseline at 24 h after dosing. Presumably, this effect is receptor-mediated since binding of AMD3100 to CXCR4 may inhibit the chemotactic effects of SDF-1α, causing release of WBC's from the endothelium, and/or progenitor cells from bone marrow, into the general circulation.[56–58] AMD3100 demonstrated dose-proportionality over the entire study range. At the highest dose level (80 μg/kg) the median C_{max} was 515 ng/mL and AUC_{inf} was 1044 ng/h/mL. Using a two-compartment model, the pharmacokinetic parameter estimates in the two highest dose cohorts were: elimination half-life of 3.6 h; volume of distribution 0.34 L/kg; clearance 1.3 L/h. The volume of distribution was larger than normal for blood volume suggesting that AMD3100 distributes freely beyond the central blood compartment and possibly into peripheral sites such as the lymph nodes, through which infected cells traffic. After a single intravenous dose of AMD3100, a drug concentration exceeding that required to inhibit HIV-1 viral replication by 90% *in vitro* was achieved for 12 h. Five volunteers also received a single subcutaneous injection of AMD3100 at either 40 or 80 μg/kg which resulted in a median systemic bioavailability of 87%. Consistent with preclinical animal models, three subjects receiving a single oral dose of up to 160 μg/kg did not show detectable drug levels in the blood.

IX. CONCLUSIONS

The identification of anti-HIV agents with new mechanisms of action remains an important therapeutic goal both to complement existing therapies and possibly impede the development of drug-resistant HIV strains. In this chapter, we have described the antiviral activity, structure, and chemistry of a new class of compounds, the bicyclams, which exhibit potent and selective inhibition of HIV replication by binding to the chemokine receptor CXCR4, the coreceptor used by T-tropic HIV for fusion and viral entry into the cell. We have also outlined our early approaches to the development of a CXCR4 antagonist that is efficiently orally absorbed. This led to the identification of AMD3329 (AMD3394), our most potent inhibitor of HIV replication reported to date. Several compounds from this class are currently undergoing oral bioavailability experiments and the results will be reported in due course.

Although there continues to be considerable debate over the appropriate choice of chemokine target(s) (CXCR4 or CCR5) for the development of chemotherapeutic agents, we believe that chemokine receptor antagonists in general represent the next generation of anti-

HIV agents. In August 1998, AMD3100 became the first chemokine antagonist to enter clinical trials. Following successful completion of a Phase I safety and tolerability study in healthy volunteers, AMD3100 entered Phase II clinical testing in May 1999. We eagerly await the results of efficacy testing in HIV-infected patients.

ACKNOWLEDGMENT

We thank Genevieve Kreye for excellent editorial assistance during the preparation of this manuscript.

REFERENCES

1. Barefield, E. K.; Cheung, D.; Van Derveer, D. G. *J. Chem. Soc., Chem. Comm.* **1981**, 302–304.
2. De Clercq, E.; Yamamoto, N.; Pauwels, R.; Baba, M.; Schols, D.; Nakashima, H.; Balzarini, J.; Debyser, Z.; Murrer, B. A.; Schwartz, D.; Thornton, D.; Bridger, G.; Fricker, S.; Henson, G.; Abrams, M.; Picker, D. *Proc. Natl. Acad. Sci. USA* **1992**, *89*, 5286–5290.
3. De Clercq, E.; Yamamoto, N.; Pauwels, R.; Balzarini, J.; Witvrouw, M.; De Vreese, K.; Debyser, Z.; Rosenwirth, B.; Peichl, P.; Datema, R.; Thornton, D.; Skerlj, R.; Gaul, F.; Padmanabhan, S.; Bridger, G.; Henson, G.; Abrams, M. *Antimicrob. Agents Chemother.* **1994**, *38(4)*, 668–674.
4. Donzella, G. A. S.; Lin, D.; Este S. W.; Nagashima, K. A.; Maddon, P. J.; Allaway, G. P.; Sakmar, T. P.; Henson, G.; De Clercq, E.; Moore, J. P. *Nature Medicine* **1998**, *4(1)*, 72–77.
5. Schols, D.; Struyf, S.; Damme, J. V.; Este, J. A.; Henson, G.; De Clercq, E. *J. Exp. Med.* **1997**, *186(8)*, 1383–1388.
6. Schols, D.; Este, J. A.; Henson, G.; De Clercq, E. *Antiviral Res.* **1997**, *35*, 147–156.
7. Labrosse, B.; Brelot, A.; Heveker, N.; Sol, N.; Schols, D.; De Clercq, E.; Alizon, M. *J. Virol.* **1998**, *72(8)*, 6381–6388.
8. Este, J. A.; Cabrere, C.; De Clercq, E.; Struyf, S.; Van Damme, J.; Bridger, G.; Skerlj, R.T.; Abrams, M.J.; Henson, G.; Gutierrez, A.; Clotet, B.; Schols, D. *Mol. Pharmacol.* **1999**, *55*, 65–73.
9. Feng, Y.; Broder, C. C.; Kennedy, P. E.; Berger, E. A. *Science* **1996**, *272*, 872–877.
10. Bleul, C. C.; Farzan, M.; Choe, H.; Parolin, C.; Clark-Lewis, I.; Sodroski, J.; Springer, T. A. *Nature* **1996**, *382*, 829–833.
11. Oberlin, E.; Amara, A.; Bachelerie, F.; Bessia, C.; Virelizier, J. L.; Arenzana-Seisdedos, F.; Schwartz, O.; Heard, J. M.; Clark-Lewis, I.; Legler, D. F.; Loetscher, M.; Baggiolini, M.; Moser, B. *Nature* **1996**, *382*, 833–835.

12. Cocchi, F.; DeVico, A. L.; Garzino-Demo, A.; Arya, S. K.; Gallo, R. C.; Lusso, P. *Science* **1995**, *270*, 1811–1815.

13. Dragic, T.; Litwin, V.; Allaway, G. P.; Martin, S. R.; Huan, Y.; Nagashima, K. A.; Cayanan, C.; Maddon, P. J.; Koup, R. A.; Moore, J. P.; Paxton, W. A. *Nature* **1996**, *381*, 667–673.

14. Deng, H.; Liu, R.; Ellmeier, W.; Choe, S.; Unutmaz, D.; Burkhart, M.; Di Marzio, P.; Marmon, S.; Sutton, R. E.; Hill, C. M.; Davis, C. B.; Peiper, S. C.; Schall, T. J.; Littman, D. R.; Landau, N. R. *Nature* **1996**, *381*, 661–666.

15. Alkhatib, G.; Combadiere, C.; Broder, C. C.; Feng, Y.; Kennedy, P. E.; Murphy, P. M.; Berger, E. A. *Science* **1996**, *272*, 1955–1958.

16. Zhang, L.; He, T.; Huang, Y.; Chen, Z.; Guo, Y.; Wu, S.; Kunstman, K. J.; Brown, R.C.; Phair, J. P.; Neumann, A. U.; Ho, D. D.; Wolinsky, S.M. *J. Virol.* **1998**, *72(11)*, 9307–9312.

17. Zhang, Y. J.; Moore, J. P. *J. Virol.* **1999**, *73(4)*, 3443–3448.

18. Simmons, G.; Reeves, J. D.; McKnight, A.; Dejucq, N.; Hibbits, S.; Power, C. A.; Aarons, E.; Schols, D.; De Clercq, E.; Proudfoot, A. E. I.; Clapham, R. P. *J. Virol.* **1998**, *72(10)*, 8453–8457.

19. Simonds, G.; Wilkinson, D.; Reeves, J. D.; Dittmar, M. T.; Beddows, S.; Weber, J.; Carnegie, G.; Desselberger, U.; Gray, P. W.; Weiss, R. A.; Clapham, P. R. *J. Virol.* **1996**, *70*, 8355–8360.

20. Berger, E. A.; Murphy, P. M.; Farber, J. M. *Annu. Rev. Immunol.* **1999**, *17*, 657–700.

21. Baggiolini, M. *Nature* **1998**, *392(9)*, 565–568.

22. Locati, M.; Murphy, P. M. *Ann. Rev. Med.* **1999**, *50*, 425–440.

23. Lapham, C. K.; Zaitseva, M. B.; Lee, S.; Romanstseva, T.; Golding, H. *Nature Medicine* **1999**, *5(3)*, 303–308.

24. Gupta, S. K.; Lysko, P. G.; Kodandaram, P.; Ohlstein, E.; Stadel, J. M. *J. Biol. Chem.* **1998**, *273(7)*, 4282–4287.

25. Volin, M. V.; Joseph, L.; Shockley, M. S.; Davies, P. F. *Biochem. Biophys. Res. Commun.* **1998**, *242*, 46–53.

26. Hesselgesser, J.; Halks-Miller, M.; DelVecchio, V.; Peiper, S. C.; Hoxie, J.; Kolson, D. L.; Taub, D.; Horuk, R. *Curr. Biol.* **1997**, *7*, 112–121.

27. Hesselgesser, J.; Horuk, R. *J. Neuro. Virol.* **1999**, *5*, 13–26.

28. Hesselgesser, J.; Taub, D.; Baskar, P.; Greenberg, M.; Hoxie, J.; Kolson, D. L.; Horuk, R. *Curr. Biol.* **1998**, *8*, 595–598.

29. Lavi, E.; Strizki, J. M.; Ulrich, A. M.; Zhang, W.; Fu, L.; Wang, Q.; O'Connor, M.; Hoxie, J. A.; Gonzalez-Scarano, F. *J. Am. Path.* **1997**, *151(1)*, 1035–1041.

30. Glabinski, A. R.; Ransohoff, R. M. *J. Neuro. Virol.* **1999**, *5*, 3–12.

31. Zaitseva, M.; Blauvelt, A.; Lee, S.; Lapham, C. K.; Klaus-Kovtun, V.; Mostowski, H.; Manischewitz, J.; Golding, H. *Nature Medicine* **1997**, *3(12)*, 1369–1375.

32. Tersmette, M.; De Goede, R. E. Y.; Al, B. J.; Winker, I.; Gruters, B.; Cuypers, H. T.; Huisman, H. G.; Miedema, F. *J. Virol.* **1988**, *62*, 2026–2032.

33. Fenyo, E. M.; Morfeldt-Manson, L.; Chiodi, F.; Lind, A.; von Gegerfelt, A.; Albert, J.; Olausson, E.; Asjo, B. *J. Virol.* **1988**, *62*, 4414–4419.
34. Connor, R. I.; Ho, D. D. *J. Virol.* **1994**, *68*, 4400–4408.
35. Schuitemaker, H.; Koot, M.; Kootstra, N. A.; Dercksen, M. W.; de Goede, R. E.; van Steenwijk, R. P.; Lange, J. M.; Schattenkerk, J. K.; Miedema., F.; Tersmette, M. *J. Virol.* **1992**, *66*, 1354–1360.
36. Bjorndal, A.; Deng, H.; Jansson, M.; Fiore, J. R.; Colognesi, C.; Karlsson, A.; Albert, J.; Scarlattie, G.; Littman, D. R.; Fenyo, E. M. *J. Virol.* **1997**, *71*, 7478–8487.
37. Zhang, L.; Huang, Y.; He, T.; Cao, Y.; Ho, D. D. *Nature* **1996**, *383*, 768.
38. Morner, A.; Bjorndal, A.; Albert, J.; Kewalramani, V. N.; Littman, D. R.; Inoue, R.; Thorstensson, R.; Fenyo, E. M.; Bjorling, E. *J. Virol.* **1999**, *73*, 2343–2349.
39. Ostrowski, M. A.; Justement, S. J.; Catanzara, A.; Hallahan, C. A.; Ehler, L. A.; Mizell, S. B.; Kumar, P. N.; Mican, J. A.; Chun, T. W.; Fauci, A. S. *J. Immun.* **1998**, *161*, 3195–3201.
40. Grivel, J. C.; Margolis, L. B. *Nature Medicine* **1999**, *5(3)*, 344–346.
41. Liu, R.; Paxton, W. A.; Choe, S.; Ceradini, D.; Martin, S. R.; Horuk, R.; MacDonald, M. E.; Stuhlmann, H.; Koup, R. A.; Landau, N. R. *Cell* **1996**, *86*, 367–377.
42. Samson, M.; Libert, F.; Doranz, B. J.; Rucker, J.; Liesnard, C.; Farber, C. M.; Saragosti, S.; Lapoumeroulie, C.; Cotgnaux, J.; Forceille, C.; Muyldermans, G.; Verhofstede, C. R.; Burtonboy, G.; Georges, M.; Imai, T.; Rana, S.; Yi, Y.; Smyth, R. J.; Collman, R. G.; Doms, R. W. *Nature* **1996**, *382*, 722–725.
43. Michael, N. L.; Chang, G.; Louie, G.; Mascola, J. R.; Dondero, D.; Birx, D. L.; Sheppard, H. W. *Nature Med.* **1997**, *3*, 338–340.
44. Michael, N. L.; Nelson, J. A. E.; KewalRamani, V. N.; Change, G.; O'Brien, S. J.; Mascola, J. R.; Volsky, B.; Louder, M.; White, G. C.; Littman, D. R.; Swanstrom, R.; O'Brien, T. R. *J. Virol.* **1998**, *72*, 6040–6047.
45. O'Brien, T.; Winker, C.; Dean, M.; Nelson, J. A.; Carrington, M.; Michael, N. L.; White, G. C. II. *Lancet* **1997**, *349*, 1219.
46. Zhang, L.; Carruthers, C. D.; He, T.; Huang, Y.; Cao, Y.; Wang, G.; Hahn, B.; Ho, D. D. *AIDS Res. Hum. Retroviruses* **1997**, *13*, 1357–1366.
47. Rana, S.; Besson, G.; Cook, D.; Rucker, J.; Smyth, R.; Yi, Y.; Turner, J.; Guo, H.; Du, J.; Peiper, S.; Lavi, E.; Samson, M.; Libert, F.; Liesnard, C.; Vassart, G.; Doms, R.; Parmentieer, M.; Collman, R. *J. Virol.* **1997**, *71*, 3219–3227.
48. Theodorou, I.; Meyer, L.; Magierowska, M.; Katlama, C.; Rouzioux, C.; and the Seroco Study Group. *Lancet* **1997**, *349*, 1219–1220.
49. Chan, S. Y.; Speck, R. F.; Power, C.; Gaffen, S. L.; Chesebro, B.; Goldsmith, M. A. *J. Virol.* **1999**, *73*, 2350–2358.
50. Wong, J.; Ignacio, C.; Torriani, F.; Havlir, D.; Fitch, N.; Richman, D. *J. Virol.* **1997**, *71*, 2059–2071.
51. Van't Wout, R. R. L.; Kuiken, C.; Kootstra, N.; Pals, S.; Schuitemaker, H. *J. Virol.* **1998**, *72*, 488–496.

52. Zou, Y. R.; Kottmann, A. H.; Kuroda, M.; Taniuchi, I.; Littman, D. R. *Nature* **1998**, *393*, 595–599.

53. Tachibana, K.; Hirota, S.; Lizasa, H.; Yoshide, H.; Kawabata, K.; Kataoka, Y.; Kitamura, Y.; Matsushima, K.; Yoshida, N.; Nishikawa, S.; Kishimoto, T.; Nagasawa, T. *Nature* **1998**, *393*, 591–594.

54. Nagasawa, T.; Hirota, S.; Tachibana, K.; Takakura, N.; Nishikawa, S.; Kitamura, Y.; Yoshida, N.; Kikutani, H.; Kishimoto, T. *Nature* **1996**, *382*, 635–638.

55. Bleul, C. C.; Schultze, J. L.; Springer, T. A. *J. Exp. Med.* **1998**, *187*, 753–762.

56. Viardot, A.; Kronenwett, R.; Deichmann, M.; Haas, R. *Ann. Hematol.* **1998**, *77*, 195–197.

57. Aiuti, A.; Webb, L. J.; Bleul, C.; Springer, T.; Gutierrez-Ramos, J. C. *J. Exp. Med.* **1997**, *185*, 111–120.

58. Peled, A.; Petit, I.; Kollet, O.; Magid, M.; Ponomaryov, T.; Byk, T.; Nagler, A.; Ben-Hur, H.; Many, A.; Shultz, L.; Lider, O.; Alon, R.; Zipori, D.; Lapidot, T. *Science* **1999**, *283*, 845–848.

59. Qing, M.; Jones, D.; Springer, T. A. *Immunity* **1999**, *10*, 463–471.

60. Louache, F.; Henri, A.; Bettaieb, A.; Oksenhendler, E.; Raguin, G.; Tuliez, M.; Vainchenker, W. *Blood* **1992**, *80*, 2991–2999.

61. Bagnara, G. P.; Zauli, G.; Giovannini, M.; Re, M. C.; Furlini, G.; La Place, M. *Exp. Hematol.* **1990**, *18*, 426–430.

62. Marandin, A.; Katz, A.; Oksenhendler, E.; Tulliez, M.; Picard, F.; Vainchenker, W.; Louache, F. *Blood* **1996**, *88*, 4568–4578.

63. Moore, B. B.; Arenberg, D. A.; Strieter, R. M. *Trends Cardiovasc. Med.* **1998**, *8(2)*, 51–57.

64. Moore, B. B.; Keane, M. P.; Addison, C. L.; Arenberg, D. A.; Strieter, R. M. *J. Investigative Med.* **1998**, *46*, 113–120.

65. Strieter, R. M.; Polverini, P. J.; Kunkel, S. L.; Arenberg, D. A.; Burdick, M. D.; Kasper, J.; Dzuiba, J.; Van Damme, J.; Walz, A.; Marriott, D.; Chan, S. Y.; Roczniak, S.; Shanafelt, A. B. *J. Biol. Chem.* **1995**, *270*, 27348–27357.

66. Mosier, D. E.; Picchio, G. R.; Gulizia, R. J.; Sabbe, R.; Poignard, P.; Picard, L.; Offord, R. E.; Thompson, D. A.; Wilken, J. *J. Virol.* **1999**, *73*, 3544–3550.

67. Este, J. A.; Cabrera, C.; Blanco, J.; Schols, D.; Gutierrez, A.; Bridger, G.; Henson, G.; Clotet, B.; Schols, D.; De Clercq, E. *J. Virol.* **1999**, *73*, 5577–5585.

68. Herbein, G.; Mahiknecht, U.; Batliwalla, F.; Gregersen, P.; Pappas, T.; Buterl, J.; O'Brien, W. A.; Verdin, E. *Nature* **1998**, *395*, 189–194.

69. Buttini, M.; Westland, C. E.; Masliah, E.; Yafeh, A. M.; Wyss-Coray, T.; Mucke, L. *Nature Medicine* **1998**, *4(4)*, 441–446.

70. De Clercq, E. *J. Med. Chem.* **1995**, *38*, 2491–2517.

71. De Vreese, K.; Reymen, D.; Griffin, P.; Steinkasserer, A.; Werner, G.; Bridger, G. J.; Este, J.; James, W.; Henson, G. W.; Desmyter, J.; Anne, J.; De Clercq, E. *Antiviral Res.* **1996**, *29*, 209–219.

72. De Vreese, K.; Kofler-Mongold, V.; Leutgeb, C.; Weber, V.; Vermeire, K.; Schacht, S.; Anne, J.; De Clercq, E.; Datema, R.; Werner, G. *J. Virol.* **1996**, *70(2)*, 689–696.

73. De Vreese, K.; Van Nerum, I.; Vermeire, K.; Anne, J.; De Clercq, E. *Antimicrob. Agents Chemother.* **1997**, *41(12)*, 2616–2620.

74. Lai, T. F.; Poon, C. K. *Inorg. Chem.* **1976**, *15(7)*, 1562–1566.

75. Doranz, B. J.; Orsini, M. J.; Turner, J. D.; Hoffman, T. L.; Berson, J. F.; Hoxie, J. A.; Peiper, S. C.; Brass, L. F.; Doms, R. W. *J. Virol.* **1999**, *73(4)*, 2752–2761.

76. Crump, M. P.; Gong, J. H.; Loetscher, P.; Rajarathnam, K.; Amara, A.; Arenzana-Seisdedos, F.; Virelizier, J. L.; Baggiolini, M.; Sykes, B. D.; Clark-Lewis, I. *EMBO J.* **1997**, *16*, 6996–7007.

77. Heveker, N.; Montes, M.; Germeroth, L.; Amara, A.; Trautmann, A.; Alizon, M.; Schneider-Mergener, J. *Current Biology* **1998**, *8*, 369–376.

78. Willett, B. J.; Adema, K.; Heveker, N.; Brelot, A.; Picard, L.; Alizon, M.; Turner, J. D.; Hoxie, J. A.; Peiper, S.; Neil, J. C.; Hosie, M. J. *J. Virol.* **1998**, *72*, 6475–6481.

79. Egberink, H. F.; De Clercq, E.; Van Vliet, L. W.; Balzarini, J.; Bridger, G.; Henson, G.; Horzinek, M. C.; Schols, D. *J. Virol.* **1999**, *73*, 6346–6352.

80. Schols, D.; Este, J. A.; Cabrera, C.; De Clercq, E. *J. Virol.* **1998**, *72(5)*, 4032–4037.

81. Hoxie, J. A.; LaBranche, C. C.; Endres, M. J.; Turner, J. D.; Berson, J. F.; Doms, R. W.; Matthews, T. J. *J. Reproductive Immunol.* **1998**, *41*, 197–211.

82. Lapham, C. K.; Ouyang, J.; Chandrasekhar, B.; Nguyen, N. Y.; Dimitrov, D. S.; Golding, H. *Science* **1996**, *274*, 602–605.

83. Kilby, J. M.; Hopkins, S.; Venetta, T. M.; DiMassimo, B.; Cloud, G. A.; Lee, J. Y.; Alldredge, L.; Hunter, E.; Lambert, D.; Bolognesi, D.; Matthew, T.; Johnson, M. R.; Nowak, M. A.; Shaw, G. M.; Saag, M. S. *Nature Medicine* **1998**, *4(11)*, 1302–1307.

84. Este, J. A.; De Vreese, K.; Witvrouw, M.; Schmit, J. C.; Vandamme, A. M.; Anne, J.; Desmyter, J.; Henson, G. W.; Bridger, G.; De Clercq, E. *Antiviral Res.* **1996**, *29*, 297–307.

85. Schols, D.; De Clercq, E. *J. Gen. Virol.* **1998**, *79*, 2203–2205.

86. Richardson, J.; Pancino, G.; Merat, R.; Leste-Lasserre, T.; Moraillon, A.; Schneider-Mergener, J.; Alizon, M.; Sonigo, P.; Heveker, N. *J. Virol.* **1999**, *73*, 3661–3671.

87. Datema, R.; Rabin, L.; Hincenbergs, M.; Moreno, M. B.; Warren, S.; Linquist, V.; Rosenwirth, B.; Seifert, J.; McCune J. M. *Antimicrob. Agents Chemother.* **1996**, *40(3)*, 750–754.

88. Bridger, G. J.; Skerlj, R. T.; Pabmanabhan, S.; Thornton, D. *J. Org. Chem.* **1996**, *62*, 1519–1522.

89. Bridger, G. J.; Skerlj, R. T.; Thornton, D.; Pabmanabhan, S.; Martelluci, S. A.; Henson, G. W.; Abrams, M. J.; Yamamoto, N.; De Vreese, K.; Pauwels, R.; De Clercq, E. *J. Med. Chem.* **1995**, *38*, 366–378.

90. Izatt, R. M.; Pawlak, K.; Bradshaw, J. S.; Bruening, R. L. *Chem. Rev.* **1991**, *91*, 1721–2085.

91. Cabbiness, D. K.; Margerum, D. W. *J. Am. Chem. Soc.* **1970**, *92*, 2151.

92. Billo, E. J. *Inorg. Chem.* **1984**, *23*, 236.

93. Barefield, E. K. *Inorg. Chem.* **1972**, *9*, 2273–2274.

94. Adam, K. R.; Antolovich, M.; Atkinson, I. M.; Leong, A. J.; Lindoy, L. F.; McCool, B. J.; Lindsey Davis, R.; Kennard, C. H. L.; Tasker, P. A. *J. Chem. Soc.* **1994**, 1539–1540.

95. Hosseini, M. W.; Lehn, J. M. *Helv. Chim. Acta* **1986**, *69*, 587–603.

96. Kimura, E.; Sakonaka, A.; Yatsunami, T.; Kodama, M. *J. Am. Chem. Soc.* **1981**, *103*, 3041–3045.

97. Nave, C.; Truter, M. R. *J. Chem. Soc., Dalton. Trans.* **1974**, 2351–2354.

98. Kajiwara, T.; Yamaguchi, T.; Kido, H.; Kawabata, S.; Kuroda, R.; Ito, T. *Inorg. Chem.* **1993**, *32*, 4990–4991.

99. Tasker, P. A.; Sklar, L. *J. Cryst. Mol. Struct.* **1975**, *5*, 329–344.

100. Connolly, P. J.; Billo, E. J. *J. Chem. Soc.* **1987**, *26*, 3224–3226.

101. Choi, H. J.; Suh, M. P. *J. Am. Chem. Soc.* **1998**, *120*, 10622–10628.

102. Shionoya, M.; Kimura, E.; Shiro, M. *J. Am. Chem. Soc.* **1993**, *115*, 6730–6737.

103. Shionoya, M.; Sugiyama, M.; Kimura, E. *J. Chem. Soc., Chem. Commun.* **1994**, 1747–1748.

104. Kimura, E.; Aoki, S.; Koike, T.; Shiro, M. *J. Am. Chem. Soc.* **1997**, *119*, 3068–3076.

105. Fujioka, H.; Koike, T.; Yamada, N.; Kimura, E. *Heterocycles* **1996**, *42(2)*, 775–787.

106. Inouye, Y.; Kanamori, T.; Yoshida, T.; Koike, T.; Shionoya, M.; Fujioka, H.; Kimura, E. *Biol. Pharm. Bull.* **1996**, *19(3)*, 456–458.

107. Sosa-Torres, M. E.; Toscano, R. A. *Acta Cryst.* **1997**, *C53*, 1585–1588.

108. Nonoyama, M.; Yamaguchi, T.; Kimijiri, H. *Polyhedron* **1986**, *5(11)*, 1885–1890.

109. Simon, E.; Haridon, P. L.; Pichon, R.; L'Her, M. *Inorg. Chim. Acta* **1998**, *282*, 173–179.

110. Sovilj, S. P.; Babic-Samardzija, K.; Minic, D. M. *J. Serb. Chem. Soc.* **1998**, *63(12)*, 979–985.

111. Poon, C. K.; Tobe, M. L. *J. Chem. Soc.* **1968**, *(A)*, 1549–1555.

112. Poon, C. K.; Tobe, M. L. *Inorg. Chem.* **1968**, *7(6)*, 2398–2404.

113. Cooksey, C. J.; Tobe, M. L. *Inorg. Chem.* **1978**, *17*, 1558–1562.

114. Bridger, G. J.; Skerlj, R. T.; Padmanabhan, S.; Martelluci, S. A.; Henson, G. W.; Abrams, M. J.; Joao, H. C.; Witvrouw, M.; De Vreese, K.; Pauwels, R.; De Clercq, E. *J. Med. Chem.* **1996**, *39*, 109–119.

115. Kimura, E. *Tetrahedron* **1992**, *48*, 6175–6217 and references therein.

116. Kaden, T. A. *Top. Curr. Chem.* **1984**, *121*, 157–179.

117. Bernhardt, P. V.; Lawrance, G. A. *Coord. Chem. Rev.* **1990**, *104*, 297–343.

118. Kimura, E.; Koike, T.; Machida, R.; Nagai, R.; Kodama, M. *Inorg. Chem.* **1984**, *23*, 4181–4188.

119. Kimura, E.; Koike, T.; Nada, H.; Iitaka, Y. *J. Chem. Soc., Chem. Commun.* **1986**, 1322–1323.

120. Kimura, E.; Kotake, Y.; Koike, T.; Shionoya, M.; Shiro, M. A. *Inorg. Chem.* **1990**, *29*, 4991–4996.

121. Pallavicini, P. S.; Perotti, A.; Poggi, A.; Seghi, B.; Fabbrizzi, L. *J. Am. Chem. Soc.* **1987**, *109*, 5139–5144.

122. Lotz, T. J.; Kaden, T. A. *J. Chem. Soc., Chem. Commun.* **1977**, 15–16.

123. Lotz, T. J.; Kaden, T. A. *Helv. Chim. Acta* **1978**, *61*, 1376–1387.

124. Alcock, N. W.; Kingston, R. G.; Moore, P.; Pierpoint, C. *J. Chem. Soc., Dalton Trans.* **1984**, 1937–1943.

125. Alcock, N. W.; Moore, P.; Pierpont, C. *J. Chem. Soc., Dalton Trans.* **1984**, 2371–2376.

126. Alcock, N. W.; Balakrishnan, K. P.; Moore, P. *J. Chem. Soc., Chem. Commun.* **1985**, 1731–1733.

127. Schiegg, A.; Kaden, T. A. *Helv. Chim. Acta* **1990**, *73*, 716–721.

128. Alcock, N. W.; Curzon, E. H.; Moore, P.; Omar, H. A. A.; Pierpoint, C. *J. Chem. Soc., Dalton Trans.* **1985**, 1361–1364.

129. Bridger, G. J.; Skerlj, R. T.; Padmanabhan, S.; Martelluci, S. A. , Henson, G. W.; Struyf, S.; Witvrouw, M.; De Vreese, K.; Schols, D.; De Clercq, E. *J. Med. Chem.* in press (1999).

130. Costa, J.; Delgado, R. *Inorg. Chem.* **1993**, *32*, 5257–5265.

131. Bradshaw, J. S.; Krakowiak, K. E.; Izatt, R. M. *The Chemistry of Heterocyclic Compounds: Aza-Crown Macrocycles, 1.* Taylor, E. C., Ed.; New York: John Wiley, 1993.

132. Krakowiak, K. E.; Bradshaw, J. S.; Zamecka-Krakowiak, D. J. *Chem. Rev.* **1989**, *89*, 929–972.

133. Ciampolini, M.; Fabbrizzi, L.; Perotti, A.; Poggi, A. N.; Seghi, B.; Zanobini, F. *Inorg. Chem.* **1987**, *26*, 3527–3533.

134. Qian, L.; Sun, Z.; Mertes, M. P.; Mertes, K. B. *J. Org. Chem.* **1991**, *56*, 4904–4907.

135. Herve, G.; Bernard, H.; Le Bris, N.; Yaouanc, J. J.; Handel, H. *Tetrahedron Lett.* **1998**, *39*, 6861–6864.

136. Weisman, G. R.; Reed, D. P. *J. Org. Chem.* **1996**, *61*, 5186–5187.

137. Reed, D. P.; Weisman, G. R.; Wong, E. H. *Abstracts of the 216th ACS National Meeting, Boston*, August 23–27, 1998.

138. Kruper, W. J.; Rudolf, P. R.; Langhoff, C. A. *J. Org. Chem.* **1993**, *58*, 3869–3876.

139. Gaudinet-Harmann, B.; Zhu, J.; Fensterbank, H.; Larpent, C. *Tetrahedron Lett.* **1999**, *40*, 287–290.

140. Richman, J. E.; Atkins, T. J. *J. Am. Chem. Soc.* **1974**, *96*, 2268–2270.

141. Atkins, T. J.; Richman, J. E.; Oettle, W. F. *Org. Synth.* **1978**, *58*, 86–98.

142. Hediger, M.; Kaden, T. A. *Helv. Chim. Acta* **1983**, *66*, 861–870.

143. Vriesema, B. K.; Butler, J.; Kellogg, R. M. *J. Org. Chem.* **1984**, *49*, 110–113.

144. Chavez, F.; Sherry, A. D. *J. Org. Chem.* **1989**, *54*, 2990–2992.

145. Kimura, E.; Shionoya, M.; Okamoto, M.; Nada, H. *J. Am. Chem. Soc.* **1988**, *110*, 3679.

146. Kimura, E.; Koike, T.; Takahashi, M. *J. Chem. Soc., Chem. Commun.* **1985**, 385–386.

147. Morphy, R. J.; Parker, D.; Alexander, R.; Bains, A.; Carne, A. F.; Eaton, M. A. W. *J. Chem. Soc., Chem. Commun.* **1988**, 156.
148. Tabushi, I.; Tanaguchi, Y.; Kato, H. *Tetrahedron Lett.* **1977**, *12*, 1049–1052.
149. Moi, M. K.; Meares, C. F.; McCall, M. J.; Cole, W. C.; DeNardo, S. J. *Anal. Biochem.* **1985**, *148*, 249.
150. Desphande, S. V.; DeNardo, S. J.; Meares, C. F.; McCall, M. J.; Adams, G. P.; Moi, M. K.; DeNardo, G. L. *J. Nucl. Med.* **1988**, *29*, 217.
151. Takenouchi, K.; Watanabe, K.; Kato, Y.; Koike, T.; Kimura, E. *J. Org. Chem.* **1993**, *58*, 1955.
152. McMurray, T. J.; Brechbiel, M.; Kumar, K.; Gansow, O. A. *Bioconjugate Chem.* **1992**, *3*, 108.
153. Wagler, T. R.; Fang, Y.; Burrows, C. J. *J. Org. Chem.* **1989**, *54*, 1584.
154. Kise, N.; Oike, H.; Okazaki, E.; Yoshimoto, M.; Shono, T. J. *J. Org. Chem.* **1995**, *60*, 3980.
155. Wagler, T. R.; Burrows, C. J. *J. Chem. Soc., Chem. Commun.* **1987**, 277.
156. Benabdallah, T.; Guglielmetti, R. *Helv. Chim. Acta* **1988**, *71*, 602.
157. Nahm, S.; Weinreb, S. M. *Tetrahedron Lett.* **1981**, *22*, 3815.
158. Sibi, M. P. *Org. Prep. Proceed. Int.* **1993**, *25*, 15.
159. Xu, D.; Mattner, P. G.; Prasad, K.; Repic, O.; Blacklock, T. J. *Tetrahedron Lett.* **1996**, *37*, 5301–5304.
160. Guillaume, D.; Marshall, G. R. *Synthetic Comm.* **1998**, *28(15)*, 2903–2906.
161. Gardinier, I.; Roignant, A.; Oget, N.; Bernard, H.; Yaouanc, J. J.; Handel, H. *Tetrahedron Lett.* **1996**, *37*, 7711.

INDEX

Printed and bound by CPI Group (UK) Ltd, Croydon, CR0 4YY

03/10/2024

01040434-0004